Winning In Asia, European Style

Market and Nonmarket Strategies for Success

Edited by Vinod K. Aggarwal

Softcover reprint of the hardcover 1st edition 2002 978-0-312-23913-8

First published 2001 by PALGRAVE™
175 Fifth Avenue, New York, N.Y. 10010 and
Houndmills, Basingstoke, Hampshire RG21 6XS.
Companies and representatives throughout the world

PALGRAVE is the new global publishing imprint of St. Martin's Press LLC
Scholarly and Reference Division and Palgrave Publishers Ltd (formerly
Macmillan Press Ltd).

ISBN 978-1-349-63239-8   ISBN 978-1-137-10722-9 (eBook)
DOI 10.1057/9781137107229

Library of Congress Cataloging-in-Publication Data
Winning in Asia, European style : market and nonmarket strategies for
success / edited
by Vinod K. Aggarwal.
    p.   cm.
    Earlier drafts of the chapter were presented at two Berkeley workshops.
    Includes bibliographical references and index.
    1. Export marketing—Europe—Management.   2. Export
marketing—Asia—Management.   3. Strategic planning—Europe.
4. Strategic planning—Asia.   5. Corporations, European—Asia—Case
studies.   6. Investments, European—Asia—Case studies.
7. Europe—Foreign economic relations—Asia.   8. Asia—Foreign
economic relations—Europe.   I. Aggarwal, Vinod K.

HF1416.6.E85 W56   2001
658.8'48—dc21   2001032195

A catalogue record for this book is available from the British Library.

Design by Letra Libre, Inc.

First edition: September 2001
10 9 8 7 6 5 4 3 2 1

Contents

Preface   v
Contributors   ix
List of Acronyms   xi

PART ONE
THEORETICAL FRAMEWORK

Chapter 1   Analyzing European Firms' Market
and Nonmarket Strategies in Asia   3
Vinod K. Aggarwal

PART TWO
THE ASIAN AND EUROPEAN MARKET
AND NONMARKET ENVIRONMENTS

Chapter 2   Europe's Trade and Foreign
Direct Investment in Asia   31
Shujiro Urata

Chapter 3   Nonmarket Strategies in Asia:
The Regional Level   59
John Ravenhill

Chapter 4   Euro-Pressure: Avenues and Strategies
for Lobbying the European Union   79
Cédric Dupont

PART THREE
CASE STUDIES

Chapter 5   From Local to Global: European Enterprise
             Software Strategies in Asia   107
                 Trevor H. Nakagawa

Chapter 6   The Fast Lane to Asia:
             European Auto Firms in China   159
                 Nick Biziouras and Beverly Crawford

Chapter 7   David and Goliath:
             Airbus vs. Boeing in Asia   187
                 William Love and Wayne Sandholtz

Chapter 8   Penetrating the Regulatory Thicket in Asia:
             Nonmarket Strategies in Banking and Insurance   225
                 Klaus Wallner

PART FOUR
CONCLUSION

Chapter 9   Lessons from European Firm Strategies in Asia   257
                 Vinod K. Aggarwal

Index   281

Preface

The Asian crises of 1997 dramatically altered the short-term attractiveness of Asian markets. The almost unbridled optimism that rapid growth would continue in the region had made these markets a central focus for European, American, and other foreign firms for many years. During the darkest moments of the crisis, however—with currencies plunging in several countries and fear of contagion growing—many foreign firms were quick to exit the market and shifted their regional strategic efforts. Yet the rapid descent of these economies was followed by an equally rapid recovery in most countries in the region. Together with opportunities to secure assets at bargain basement prices and pressures for liberalization of markets, many firms once again rushed to Asian markets.

This book analyzes how European firms have attempted to win in Asian markets, both before and after the Asian crises. A central focus of this volume is the formulation and implementation of market and nonmarket strategies. Market strategies have been the topic of many works. Yet as the case studies in the book readily demonstrate, the most successful firms have won in Asia by integrating market strategies with nonmarket strategies that help firms to respond to and benefit from shifts in the political-economic-social environment. Firms that have been able to leverage their capabilities to secure assistance from the European Union and European governments and who have developed strategic relationships with Asian governments and firms, have repeatedly emerged as winners. By examining integrated market and nonmarket strategy, both from an analytical and empirical perspective, I hope that this book will enhance our understanding of firm strategies in Asia that will be of benefit to both analysts and practitioners.

This project began as a grant application submitted to the Institute of European Studies of the University of California, just a few months before the onset of the Asian crises. I am greatly indebted to the Institute's director, Gerald Feldman, and its deputy director, Beverly Crawford, for their willingness to bet on the merits of this project before the dramatic changes in Asia made this a particularly salient issue. The staff of the Center, particularly Gia White, facilitated the smooth management of the project. The Institute's generosity encouraged the broad participation of an international group of scholars and helped us turn an idea into a book.

At two Berkeley workshops where earlier drafts of the chapters were presented, Nicolas Jabko, Chung Lee, Seungjoo Lee, Kun-Chin Lin, Greg Linden, David Moon, Elliot Posner, and Nick Ziegler served ably as discussants and stimulated valuable revisions of the papers. The contributions of other readers are acknowledged by the authors of each of the chapters.

I am especially indebted to the staff of the Berkeley APEC Study Center (BASC). At BASC, Trevor Nakagawa, Kun-Chin Lin, Zachary Zwald, Ralph Espach, Ed Fogarty, and Chris Tucker have provided valuable research assistance, comments, and helped in organizing the Berkeley workshops. In particular, Trevor Nakagawa played a key role from beginning to end in the evolution of this project. A number of undergraduates also helped in preparing the manuscript and in editing papers. For their help, I am particularly grateful to Lily Bradley, Mary Eddy, Faisal Ghori, Justin Kolbeck and Mytoan Nguyen who currently work at BASC as part of the Berkeley Undergraduate Research Apprenticeship program. Other undergraduates who have worked on one facet or another of this project include Moonhawk Kim, Catherine Oliver, Grace Wang, Brandon Yu, Brandon Loew, Niklas Ponnert, Mike Hunzeker, Rishi Chandra, and Deanna Wu.

Karen Wolny of Palgrave (formerly St. Martin's Press Scholarly and Reference Division) has been a strong supporter of my work on Asia, having published an earlier co-edited book, *Asia Pacific Crossroads*. I was so pleased by her editorial guidance and the quality of the press, that I only sought out Palgrave when I was looking for an outlet for this work. Gabriella Pearce, who works with Karen, has been of immense help in managing the publication process.

On a personal note, I would like to thank my family for their support. My parents, Om and Saroj Aggarwal, have taken on many tasks that would otherwise have distracted me from my academic work. Sonia, my ten-year old daughter, has been of great help as a research assistant and paper flow manager. As she becomes a budding writer in her own right, I am pleased to have someone to commiserate with as we both endlessly edit our respective writings. I have been especially fortunate to have the encouragement of my wife, Nibha Aggarwal, who has been exceptionally busy devising her own corporate strategy as founder of a new company, SkyFlow. I hope that this book on strategy will soon be relevant to her as SkyFlow expands to capture Asian markets.

Vinod K. Aggarwal
Berkeley, California
July 2001

Contributors

VINOD K. AGGARWAL is a Professor in the Department of Political Science, Affiliated Professor in the Haas School of Business, and Director of the Berkeley APEC Study Center (BASC) at the University of California, Berkeley.

NICK BIZIOURAS is a Ph.D. candidate at the Department of Political Science at UC Berkeley.

BEVERLY CRAWFORD is the Deputy Director of the Institute for European Studies at UC Berkeley.

CÉDRIC DUPONT is an Assistant Professor at the Graduate Institute of International Studies, Geneva.

WILLIAM LOVE is a Ph.D. candidate at the Department of Political Science at UC Irvine.

TREVOR NAKAGAWA is a BASC Project Director at UC Berkeley, and a Ph.D. candidate at the Department of Political Science at UC Berkeley.

JOHN RAVENHILL is a Professor at the University of Edinburgh.

WAYNE SANDHOLTZ is an Associate Professor of Political Science at UC Irvine.

SHUJIRO URATA is a Professor of Economics at Waseda University, Tokyo.

KLAUS WALLNER is an Assistant Professor of Agricultural and Resource Economics at Oregon State University.

List of Acronyms

4GL	fourth generation languages
ABAC	APEC Business Advisory Council
ACEA	Association of European Automobile Constructors
AFTA	ASEAN Free Trade Area
AI	Airbus Industrie
AIA	ASEAN Investment Area
AIC	ASEAN Industrial Complementation
AIG	American International Group
AIJV	ASEAN Industrial Joint Venture Agreement
ANA	All Nippon Airways
APEC	Asia-Pacific Economic Cooperation
APPE	Association of Petrochemicals Producers
ASEAN	Association of South East Asian Nations
ASEAN-CCI	ASEAN Chambers of Commerce and Industry
ASEM	Asia-Europe Meeting
B2B	business-to-business
BAe	British Aerospace
BBC	Brand-to-Brand Complementation
BC-Net	Business Cooperation Network
BIS	Bank of International Settlements
BRE	Bureau de Rapprochement des Entreprises
BRITE	Basic Research in Industrial Technologies for Europe
CAAC	Civil Aviation Authority of China
CASA	Construcciones Aeronaúticas
CASC	China Aviation Supply Corporation
CASE	Computer-Aided Software Engineering
CCITT	Comité Consultatif International Téléphonique et Télégraphique
CCMC	Community of European Community Automobiles Makers

CCP	Chinese Communist Party
CEFIC	European Chemical Industry Council
CEPT	Common External Preferential Tariff
CER	Closer Economic Relations (Australia and New Zealand)
CITIC	China International Trust and Investment Corporation
COBOL	Common Business-Oriented Language
COREPER	Committee of Permanent Representatives
DASA	Daimler Chrysler Aerospace
DGI	Directorate-General for External Economic Relations
DGIA	Subdivision of DGI
DGIB	Subdivision of DGI
DGIII	Directorate-General for Internal Market and Industrial Affairs
DGXII	Directorate-General for Science, Research and Development
DGXIII	Directorate-General for Telecommunications, Information Technology, and Industries
DGXV	Directorate-General Financial Institutions and Company Law
DGXVII	Dirctorate-General Energy
DGXXIII	Directorate-General for Enterprise Policy, Distributive Trades, Tourism, and Cooperatives
DM	Deutsche marks
EC	European Community
ECB	European Central Bank
ECIP	European Community Investment Partners
ECIS	European Community for Interoperable Systems
ECOSOC	Economic and Social Committee
ECJ	European Court of Justice
ECTEL	European Conference of Associations of Telecommunications
EEC	European Economic Community
EIB	European Investment Bank
EMU	European Monetary Union
ERP	enterprise resource planning
ERT	European Round Table of Industrialists

ESPRIT	European Strategic Program for Research and Development in Information Technology
ESSI	European Systems and Software Initiative
ETOPS	Extended Twin Overwater Operations
EU	European Union
EUROBIT	European Association of Manufacturers of Business Machines and Information Industry
EUROPACABLE	European Associations of Manufacturers of Insulated Wires and Cables
FAA	Federal Aviation Administration
FDI	foreign direct investment
Ffr	French francs
G-5	Group of Five industrial countries
GATT	General Agreement on Tariffs and Trade
GDI	gasoline direct injection
GDP	gross domestic product
GIE	Groupement d'Intérêt Économique
GM	General Motors
GUI	graphical user interface
IATA	International Air Transport Association
IBC	Integrated Broadband Communications system
IMF	International Monetary Fund
IPAP	Investment Promotion Action Plan
IPN	international production networks
ISV	independent software vendor
IT	information technology
ITA	International Technology Agreements
ITU	International Telecommunication Union
JAL	Japan Airlines
KAL	Korean Airlines
LDP	Liberal Democratic Party (Japan)
MAS	Malaysia Airline Systems
MERCOSUR	Mercado del Sur
MBB	Messerschmitt-Beolkow-Blohm
MD	McDonnell Douglas
MFA	Multi-Fiber Arrangement
MNC	multinational corporation
MoF	Ministry of Finance
MOU	memorandum of understanding

NAFTA	North American Free Trade Agreement
NIC	newly industrialized country
NIE	newly industrialized economy
ODA	Official Development Assistance
ODM	original design manufacturer
OECD	Organization for Economic Cooperation and Development
OEMs	Original Equipment Manufacturers
PTA	preferential trading arrangements
R&D	research and development
R&TD	Research and Technological Development policy
RACE	Research and Development in Advanced Communications in Europe
SAGE	Software Action Group for Europe
SAP	Systems, Applications, Products
SEA	Single European Act
SEM	Single European Market
SIA	Singapore Airlines
SMEs	small and medium-sized enterprises
SPECS	Specification and Programming Environment for Communication Software
SWOT	Strengths, Weaknesses, Opportunities, and Threat
TDA	Toa Domestic Airlines
TEU	Treaty of the European Union (Maastricht Treaty)
TFAP	Trade Facilitation Action Plan
TRIMs	Trade-Related Investment Measures
UNCTAD	United Nations Conference on Trade and Development
UNICE	Union of Industrial and Employers' Confederation
VW	Volkswagen
WIPO	World Intellectual Property Organization
WTO	World Trade Organization
XII	Directorate-General for Science, Research, and Development

Part One

Theoretical Framework

Chapter 1

Analyzing European Firms' Market and Nonmarket Strategies in Asia

Vinod K. Aggarwal[1]

I. Introduction

Despite recent currency crises, most of the Asia-Pacific economies continue to be among the most attractive markets in the world and now appear to be recovering rapidly. The previous track record of the newly industrializing countries, phenomenal Chinese growth rates, and widespread economic liberalization testify to this recovery. But the ups and downs of the Asian market have forced Japanese, American, and European firms to rethink their strategies. Some firms have responded by increasing investments in the region, hoping to benefit from a quick economic recovery and the sale of distressed assets that will leave them well positioned to profit from renewed growth in the region. Other firms are concerned that their excessive reliance on exports to Asia has made them highly vulnerable. As a result, they have sought to diversify their marketing effort and have attempted to position themselves in newly emerging markets in Latin America, Eastern Europe, and elsewhere. The key focus of this book is to analyze the strategic interplay between governments and firms in Asia. By systematically examining the nature of European investment and trade strategies in Asian markets in a variety of sectors, and by comparing European firms with American and Japanese firms (in two companion volumes), we hope to understand the factors that affect competitive success.

An important element of understanding firm strategies in Asia is the nature of nonmarket strategies.[2] Although firms must pursue market strategies to position themselves in a changing global economy, they also interact with governments to secure policies favorable to them. Firms are interested in securing access to closed or restricted markets for exports and investment, are concerned about regulations on their subsidiaries, and are wary of changing taxation policy, among other issues. In attempting to influence policy outcomes, they often work with both their home and host governments to implement policy changes. And at the same time, home and host governments have objectives of their own vis-à-vis both their own and foreign firms. In addition to understanding the strategies employed by European firms, we hope to shed light on two key questions: Do firms' market and "nonmarket" strategies vary more by industrial sector or by national origin? And how do different governments react to the push and pull from firms of different national stripes?

Our focus on Asia is driven by four key factors. First, East Asian countries provide examples of both extremely high growth rates and markets in severe recession, accompanied by International Monetary Fund (IMF) and U.S. pressures for liberalization. As the region recovers from the 1997–98 financial crisis, East Asia provides an excellent laboratory to analyze shifting firm strategies in times of good and bad fortune. Second, many Asian firms pose a significant competitive challenge to foreign firms. Not only do they often have dominant positions in their home markets, but they have been successful in entering European and American markets. Third, many Asian firms have close ties to governments. Indeed, the nature of government-business relations is particularly intricate in the Asian context. Most of the newly industrializing countries (NICs), both the so-called first and second tier, have actively used industrial policy measures in an effort to bolster their firms. Restrictions on investments, technology transfer, export performance requirements, preferential financing, and a host of other instruments have been commonplace in most of these countries. Fourth, the Asia-Pacific has been one of the most interesting arenas in the world to understand the interplay of different forms of governance in terms of regionalism, sectoralism, and globalism. This combination is nicely illustrated by the evolution of the recent Information Technology Agreement (ITA). Although this agreement to liberalize trade in a host of information technology products was initially vetted in the Quad group of countries, it was promoted actively on a sectoral basis in the regional grouping known as Asia-Pacific Economic Cooperation (APEC). It was then globalized in 1996 at the Singapore World Trade Organization (WTO) ministerial meeting and has been accepted by most countries in the world.

This chapter discusses the analytical framework and theoretical approaches that form the foundation of the analysis in our empirical chapters. The chapter is divided as follows. Section II examines the nature of *positional analysis* and how market forces, firm competencies, and the nonmarket environment influence the choice of trade, investment, or some mix, at the national, regional, or global level. In Section III, we turn to *strategic analysis,* consisting of a firm's choices of market strategy, a transaction cost analysis of organization forms of market penetration, and a distributive politics analysis of nonmarket issues. These factors combine to influence the firm's integrated strategic choice. Implementation of this choice based on *tactical analysis* is the topic of Section IV. This section considers the market, organizational, and nonmarket tactics that firms must pursue to succeed with their chosen strategy. Section V reviews the organization of the book and examines the questions explored by the case studies of the software, auto, commercial aircraft, banking, and insurance industries.

Figure 1.1 provides a roadmap of the analysis that follows in Sections II through IV.

As indicated in this figure, one can conceptualize the choice of trade or investment, integrated strategic choice, and implementation efforts based on the three "triangles" of positional, strategic, and tactical analysis. In turn, each of these triangles illuminate essential elements that firms must take into account in their analysis. Finally, based on the policies that firms actually pursue (and the policies that potential competitors pursue), a new cycle of analysis may be set in motion owing to feedback from their actions.

II. Positional Analysis: Market Factors, Core Competencies, and The Nonmarket Environment in Diverse Geographical Arenas

Firms operate in both a market and a nonmarket environment. Corporate strategists have traditionally focused on the market environment within which firms operate, and on the organization of firms. With respect to market analysis, the emphasis has been on such elements as the technological profile of the industry, the number of major players, the barriers to entry, and so forth. In addition to such market analysis, work in corporate strategy and organization has concerned itself with the way firms might be appropriately structured internally to enhance their competitive position. This research includes analysis of the development of different types of firm organization, the design of incentive systems, and so on.

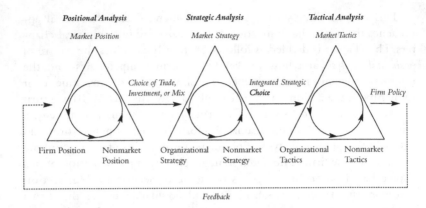

Figure 1.1   An Overview of Integrated Strategy: Triangulating Strategic Responses to Market and Nomarket Forces

Aside from these two crucial aspects to explain the performance of firms, we must pay attention to the social, political, and legal context within which they operate—in short, their nonmarket environment. With respect to nonmarket forces, the relevant consideration for our purposes is an analysis of issues, interests involved, information, and institutions,[3] particularly based on the firm's positioning, be it at the national, regional, or global level. Thus, as firms initially choose whether or not to enter Asian markets, increase their investments, or alter their trading patterns, they must consider the nonmarket issues involved in their choice with respect to specific countries. At the same time, regional and global issues may affect their strategic initiatives, and the relevant issues and institutions must be incorporated into their analysis.

The elements that we have very briefly considered provide the basis for the positional analysis "triangle" of factors depicted in figure 1.2, which we shall analyze in more detail.

As illustrated in figure 1.2, the positional analysis gives firms the initial broad choice of whether to proceed with trade, investment, or some mixture of these in Asia. Before we consider the various nodes of this triangle in more detail, we consider the importance of geographical arenas.

Geographical Orientation

Whether one undertakes market, firm, or nonmarket analysis, a prior question concerns the appropriate geographical arena. Firms must, of course,

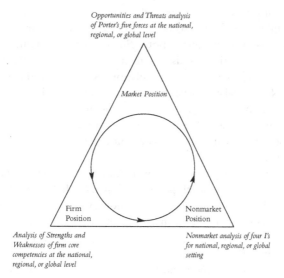

Figure 1.2   Positional Analysis

focus on the specific country's market and nonmarket characteristics that they plan to enter. This "multidomestic" focus suggests that firms need to consider the individual characteristics of different countries in their analysis.[4] At the market level, this involves a consideration of competitors, suppliers, and the like. At the nonmarket level, the types of policies that can and have been used by Asian and other countries include such things as joint venture requirements, export performance demands, local content rules, technology transfer agreements, and multilateral investment initiatives. In addition, historical political bargains and cultural values entrenched in coalitions in certain countries also affect the environment in which firms interact. Each of these factors must be dealt with on both a market and nonmarket basis.

Increasingly, however, firms must consider not only national arenas of competition, but regional and global ones as well. While the thinking on global corporate positioning is more developed, analysis of regional strategies, both from a market and nonmarket perspective, has been given short shrift. From a nonmarket perspective, in a world of growing regionalism that includes regions such as the Association of Southeast Asian Nations (ASEAN), APEC, and the European Union (EU), such developments are often accompanied by region-wide policies. In the most advanced area of integration, the movement toward a true single European market forced merger of many firms in Europe and altered market calculations. At the same time, European firms also began to develop a lobbying apparatus as many aspects of policymaking, both Europe-wide and internationally, shifted to the European Commission in Brussels. In the Asia Pacific, APEC, ASEAN, and the Closer Economic Relations (CER) accord between Australia and New Zealand have been arenas of firm influence.[5] These arrangements impact corporate strategies and influence optimal policies. This development means that firms cannot simply focus only on policymaking in specific Asian countries and ignore the nature of policymaking at the regional level.

From a market perspective, regional strategies can be crucial. Thus, if production networks are developed on an Asian-wide basis with the intent of targeting global markets, one must undertake appropriate analysis of possible competitors, suppliers, financing, and the like. Or, in trade, firms might choose a global strategy, and incorporate Asia as they would any other region into their strategy.

In terms of trends at the regional level, we can think of two potential policy changes that would influence trade and investment strategies: widening and deepening. The first of these, widening, refers to the accession of new members into existing arrangements. Deepening concerns more intense integration efforts in terms of greater coordination of monetary, social, labor, and even foreign policies. These can include such trade elements as regional content requirements, regional patent protection, regional lobbies, and the like. Both widening and deepening would of course also influence analysis of market strategies at the regional level.

Firms can, of course, concentrate on being globally oriented and competitive. At the global nonmarket level, the arrangements we have seen in the General Agreement on Tariffs and Trade (GATT) and its successor, the WTO, have greatly influenced firm strategies. For example, liberalization of specific sectors through the GATT has led to considerable competition for firms and forced them to become more internationally competitive. This includes both the reduction of tariffs and quotas as well as the removal of other kinds of nontariff barriers. In sectors such as the aircraft industry, agriculture, steel, electronics, financial services, and the like, firms must take into account new regulations and changes in the WTO. The Uruguay Round introduced a host of new issues that affect firms including changes in intellectual property protection, and the linkage between trade and investment through the Trade Related Investment Measures (TRIMS) Agreement. The agenda of the GATT and WTO has, of course, been driven by firms who have lobbied their governments with specific concerns. Thus, for example,

| | | Target Market for Sales | | |
		National	Regional	Global
Trade or	National	(1) Domestic	(2)	(3)
Investment	Regional	(4)	(5)	(6)
Location	Global	(7)	(8)	(9) Pure global

Figure 1.3   Geographical Arenas: Location and Target Markets

financial service firms in the United States were instrumental in putting the issue of financial sector liberalization on the GATT Uruguay Round agenda in 1986.[6] And many information technology firms, the entertainment industry, and pharmaceutical companies actively sought to develop a set of regulations to protect their intellectual property.

At the global sectoral level, arrangements such as the Multi-fiber Arrangement in textiles and apparel or steel voluntary export restraints have long influenced sourcing and production decisions. These arrangements have coexisted uneasily with the GATT and pressure has built to eliminate such sectoral arrangements. The new trend at this level is to open markets at the sectoral level. As noted earlier, the ITA was actively promoted in APEC and then became a multilateral agreement at the WTO December 1996 ministerial meeting. The agreement calls for the phasing out of tariffs on six major categories of equipment, namely computers, some telecommunications equipment, semiconductors, semiconductor manufacturing equipment, software, and scientific instruments. It does not include consumer electronics. In Vancouver at the November 1997 APEC meeting, following up on the information technology agreement, ministers agreed to consider nine additional sectors for fast-track trade barrier reduction: chemicals, energy-related equipment and services, environmental goods and services, forest products, medical equipment, telecommunications equipment, fish and fish products, toys, and gems and jewelry. Six remaining sectors—automotive goods, civil aircraft, food, natural and synthetic rubber, fertilizers, and oilseed products— were to be further reviewed for action in 1998. Although firms actively lobbied on all sides of this issue to advance their interests, the 1998 Kuala Lumpur APEC meeting saw a failure to advance this agenda due to Japanese resistance to liberalizing forestry and fishery products. The whole package of nine sectors has since been shifted to the WTO for negotiations.

In thinking about geographical arenas, it is useful to distinguish production from marketing orientations, both on a market and nonmarket basis. To graphically illustrate the possibilities, we can briefly consider the nine cells in figure 1.3, with two extreme points labeled to provide some bearings on strategies.

Thus, for example, one could invest in China, and simply sell nationally. Or, one could sell throughout Asia, or globally. Or alternatively, one could invest or set up on a regional basis in several countries in Asia through a trading company or production hub, and then sell only in a single country, to the whole region, or worldwide. Finally, globally-based firms could focus on single countries as well, a region, or in the "ultimate" globalization, be "pure global firm." Firms must make such choices about their location strategy based on consideration of market forces, their core competencies, and the nonmarket environment. We next turn to consideration of these elements in seriatim.

Market Forces

The most popular market analytical approach is one developed by Michael Porter based on the vast literature in industrial organization.[7] Based on ideas in this field, Porter proposed five specific factors or the "five forces model." The forces he refers to are (1) the rivalry among established firms; (2) risk of entry by potential competitors; (3) possible threat of substitutes; (4) bargaining power of suppliers; and (5) bargaining power of buyers. The forces also provide a basis for analyzing what firms face in terms of strategy formulation. Drawing on the second half of the well-known SWOT acronym (strengths, weaknesses, opportunities, and threats), market analysis examines the opportunities and threats posed by the five forces.[8]

The notion of rivalry among firms refers to the classic question of market structure, that is, whether atomistic, oligopolistic, duopolistic, or monopolistic. Varying structures directly affect whether firms can pursue autonomous strategies or must take into account the reaction of other firms because of the mutual dependence among actors in the market. The other two aspects within the concept of rivalry are demand conditions and barriers to exit. The first of these refers to the growth potential of the industry while the second concerns the impediments for firms to leave the industry. With greater growth potential, rivalry will be less intense since the game is not zero-sum. Exit barriers help to explain how firms can be resistant to exit because of high costs. This factor can also explain why firms might be more willing to take political action to resist foreign competitors.[9]

The analysis of potential competitors is based on barriers to entry. These barriers include such factors as existing brand loyalty, cost advantages arising from production techniques, and economies of scale that arise from large-scale production.[10] Other factors include the need for high capital investments, the cost of switching to another product, and access to distribution channels. Each of these barriers poses an obstacle to entry, but over time, these barriers have eroded. For example, in the steel industry, new technologies have allowed minimills to enter the market; in addition, governments may help their own firms overcome barriers to entry by subsidizing their initial efforts.

The third factor—the threat of substitutes—is straightforward. With few substitutes, firms in an industry will face little competition from outsiders. Finally, the fourth and fifth factors—bargaining power of buyers and suppliers—is part of the downstream and upstream game of market power. If buyers or suppliers are few in number, their oligopolistic position will allow them to secure better prices when interacting with firms in a particular industry.

As noted, each of these five forces can be analyzed in terms of the opportunities and threats that they pose. Most simply, the stronger the forces (competitive market structure, low barriers to entry, easy substitutes, and buyers and supplies with market power), the greater the challenge that firms face in a particular industry.

Firm Competencies

Much has been written about the elements that one must examine to understand a firm's competitive ability. Our focus in this book is primarily on external factors of markets and nonmarket environments, rather than corporate organization per se. The most popular division is between a firm's resources and its capabilities.[11] Resources, as used by many authors, focus on both tangible and intangible factors. These include such things as buildings, plant and equipment, and more intangibly, reputation, know-how, patents and the like. Capabilities in this context refer to the ability of firms to use resources in a systematic way to advance their interests, based on how the firm is structured and its control systems.

In terms of analysis, the focus is on considering the firm's strengths and weaknesses (part of the SWOT notion). Yet there is considerable debate as to what resources and capabilities constitute strengths (and the obverse, of course, weaknesses). Thus, management consultants and business school analysts have attempted to redirect attention away from the actual products that firms produce to focus on their capabilities and competencies. The most popular work on core competencies, by Gary Hamel and C. K. Prahalad, examines firms in terms of basic sets of competencies that they have developed that might then be transferred to other areas and products. Thus, rather than focus on specific resources, core competencies focus on a more complex set of capabilities that include "communication, involvement, and a deep commitment to working across organizational boundaries."[12] Using these core competencies, they argue that firms must then go on to develop core products and organize their business accordingly. This view contrasts with the focus on products made by single business units within an organization that operate in semi-autonomous manner.

Much debate rages in the literature on firm-level abilities, but the basic view is of the firm as capable of managing structural constraints systematically—rather than being at the mercy of Porter's five forces. Indeed, the literature on corporate strategy has evolved from a rather static picture of firms attempting to "fit" into the environment within which they are operating to a more dynamic perspective on how firms generate and create market opportunities for themselves.

This line of argument is best represented by the work of Prahalad and Hamel who speak of strategic "intent"—as opposed to strategic fit.[13] Thus, in their view, firms draw on their resources and capabilities to affect their market environment to dynamically position themselves to enhance their profit potential. To this, we must add the need for firms to possibly manipulate the nonmarket environment in which they operate to advantage as well.

Nonmarket Environment

Just as firms must consider their market environment and analyze the threats and opportunities, so too must they be concerned about their nonmarket environment. Specifically, as David Baron has argued, one must understand the issues that are likely to be raised in this context, the interests of major groups, the institutional setting within which policy resolution takes place, and the information available to actors.[14]

Issues can include both market-related questions as well as nonmarket problems that may impact on market activities. In an international context, and particularly in the Asian context, issues such as the environment and labor standards immediately raise potential nonmarket problems that can impact a firm's market strategy. Actors strategically interact on these issues in various institutional settings. In terms of negotiations and bargaining, it is also useful to think about the notion of issue-packaging or issue areas. These packages are linked issues and may reflect either a knowledge basis or power basis. Understanding the basis of issue linkage is crucial to analysis and the formulation of strategy.[15]

With respect to institutions and actors, whereas many analysts take a highly pluralist view of government-business relations, with nonstate actors simply vying for the attention of the government, more sophisticated understandings of the relationship between state and societal actors focus on the interests of state actors. Thus, in this analysis, institutions are not simply arenas where political activity by firms and other actors take place, with institutions as mere receptacles of interaction. In such a view, and one that is pursued in this book, the motivations and capabilities of state actors form an essential part of nonmarket analysis and strategy.

The last factor, information, refers to what is known about issues. The word "information" as used by Baron is potentially misleading, because the most significant question in understanding the nonmarket environment of issues and issue packagings is more aptly characterized as "knowledge," with implications of theoretical and causal understanding, rather than just an accumulation of facts. Knowledge, in this context, provides a key basis for the formulation of policies and will also impact the evolution of institutions. From a strategy perspective, the creation of new knowledge may provide a basis for cognitive agreement among different groups to go beyond zero-sum conflict.

Positional Analysis and the Choice of Trade and/or Investment in Asia

Our analysis to this point, then, provides a way of tapping into the broad choice of whether firms will choose to enter or increase their presence in Asia through a trade or investment strategy, or some combination of the two. The choices that countries will make on this dimension cannot be foreseen in general without a specific analysis of the market and nonmarket environment of the industry in question and the position of the firm in that industry. But as a general observation, one can note that during the Asian crisis, markets in several countries in East Asia—but by no means all—were characterized by weakening domestic firms that provided obvious targets of opportunities. This made an investment strategy more attractive than one based on increasing trade to countries in recession. Moreover, with exchange rates that were highly favorable to foreign investors, it is hardly surprising that many firms in a variety of industries chose to increase their presence in Asia. Additional impetus for investment (but to an extent, for trade as well), came from the need for countries in Asia to follow IMF policies that called for reductions in barriers to both trade and investment.

By contrast, in a non-crisis situation, both before and after the 1997–99 problems, the determinants of a trade or investment strategy would not be so clearly in favor of investment. Under "normal" conditions, then, the choice of such strategy would involve a more detailed analysis of the firm core competencies, as well as the market and nonmarket environment for specific industries. These tasks from an empirical standpoint are taken up in the case study chapters.

III. Strategic Analysis:
Market, Organization, and Nonmarket Elements

Choosing to focus primarily on trade or finance—based on an integrated consideration of market forces, firm core competencies, and the nonmarket environment—provides a first cut into the ultimate strategic choice of a firm's policy toward the Asian market. Firms must, however, face several other issues: (1) what should be the nature of a firm's market strategy with respect to product cost and quality, what technology should be transferred, and which specific market segments might one enter? (2) how might a firm decide what type of organization of trade or investment activities to pursue? and (3) what types of opposition or support will the chosen strategy receive from nonmarket actors? Figure 1.4 depicts the components making up the "strategic analysis" triangle. In each case, we can employ analytical tools to conduct strategic analysis on these various dimensions.

Market Strategy and Hypercompetition

To examine strategic choices in markets, we can draw on the work of Richard D'Aveni, who discusses the transformation of markets to one of hypercompetition.[16] In this view, firms compete in four different arenas: (1) cost and quality; (2) timing and know-how; (3) strongholds; and (4) deep pockets. In traditional perspectives, firms position themselves in one of these—say in the cost and quality arena—and might attempt to secure a high-cost/high-quality position. But as D'Aveni argues, all such static positioning efforts are ultimately (and now, more quickly) futile. Thus, firms must continually reposition themselves not only *within* arenas as the mar-

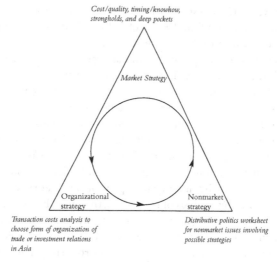

Figure 1.4   Strategic Analysis

ket evolves, but also must be continually prepared to move to compete vigorously in *different* arenas.

The first arena of competition is one in which firms compete on the basis of cost and quality. In this arena, in an "ideal typical characterization," firms initially begin with a homogeneous product and compete primarily on the basis of price. As price wars escalate, however, firms begin seeking other means of competition. Eventually, each differentiates itself from its competitors using new dimensions of quality and service. Although some firms try to cover the entire market by offering high-priced and high-quality products as well as low-priced and low-quality products, new competitors still have room to enter the market at the high or low end with either niche or outflanking strategies.

The second arena of competition, timing, and know-how may be one that firms enter or create in order to escape the unending cycle of price-quality competition present in the first arena. In this situation, a first mover can seize control of the market, but this often requires large investment in a product that can easily be imitated by competitors. To prevent imitation and maintain its control of the market, the first mover often creates barriers to market entry and develops its product in such a way as to make imitation difficult. Eventually, however, competitors do succeed in entering the market and learn to imitate the first mover's product. In response to this, the first mover may use a strategy of leapfrogging innovations in which new products are developed from large technological advances, entirely new resources, and know-how. This makes it more difficult for imitators to succeed in the market, but they do eventually catch up to the leader. The first mover will then seek a new leapfrog move, and the cycle begins again. In D'Aveni's formulation, it continues until the "next generation leapfrog strategy" is too costly and the cycle thus becomes unsustainable.[17]

In the third arena of competition, strongholds, firms seek an advantage on a playing field that may already have been leveled by price-quality competition and rapid innovation. They do this by creating strongholds to exclude competitors from their regional, industrial, or product market segments. As discussed by industrial organization theorists generally and Michael Porter in his analysis of five forces, the entry barriers that they create serve to insulate them from the price-quality and innovation-imitation cycles. Yet in contrast to this somewhat static view of barriers, in hypercompetition, such barriers provide only short-term relief, and are rarely sustainable in the long run. Competitors are likely to build war chests in their own strongholds and then fund their entry into the strongholds of others. Usually, the attacked firm will respond by defending itself and then counterattacking in the initiating firms' stronghold. In the long run, these attacks and counterattacks weaken the stronghold of both firms until no stronghold remains.

In the fourth arena of competition, firms turn to their deep pockets for an advantage. Essentially, the firms with the greatest financial resources try to gain an advantage by bullying their smaller competitors. Such bullying often includes wearing down and undercutting smaller competitors, who have fewer financial resources and therefore cannot endure in the market as long as the deep-pocketed firm. In response, the smaller competitors may develop formal or informal alliances, turn to the government for help, or step aside to avoid competition with the deep-pocketed firm. Eventually, after a series of moves and countermoves, the deep-pocketed firm has exhausted its resources and its deep-pocket advantage is either completely neutralized or substantially diminished.

Organizational Strategy

In attempting to assess how firms are likely to organize their investment or trading activities, we can draw upon a well-developed literature on transaction costs.[18] In examining contracts and organizational forms, Oliver Williamson emphasizes the importance of bounded rationality, oppor-

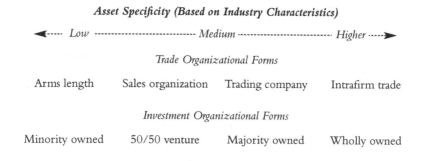

Asset Specificity (Based on Industry Characteristics)

◄----- *Low* ---------------------------- *Medium* ------------------------- *Higher* -----►

Trade Organizational Forms

Arms length   Sales organization   Trading company   Intrafirm trade

Investment Organizational Forms

Minority owned   50/50 venture   Majority owned   Wholly owned

Figure 1.5   Choosing Organizational Forms Based on Transaction Cost Analysis

tunistic behavior by actors, and the problem of highly specific assets to construct predictions about governance structures. The fundamental problem that he notes is that in executing contracts in the context of bounded rationality and opportunism, one cannot be sure if one's counterpart will perform as promised. In such cases, if a firm undertakes investments in highly specific assets, it faces a high potential probability of being exploited because these assets cannot easily be transferred to other economic activities without substantial loss.

These ideas on transaction cost problems have been developed by Witold Henisz in recent work to examine how firms might organize their foreign investment activities. Focusing on the ideas of both the importance of economic and political transaction costs, Henisz shows the interaction of contractual and political hazards.[19] Specifically, drawing on his own and others' work, he argues that firms are likely to choose majority controlled plants where contractual hazards exist. These contractual hazards include a high need to invest in specific assets, a concern that technology might be inappropriately used or exploited by a joint venture partner, and free-riding on brand-name or reputation.[20] By contrast, in the face of political hazards that include a fear of takeover by a host government, firms are likely to prefer minority investment stakes where they might be able to use the skills and political standing of their venture partners to mitigate such hazards. The interaction effect of contractual and political hazards turns out to be empirically interesting. As Henisz argues and shows empirically, when both contractual and political hazards are high, firms prefer majority owned subsidiaries because their joint venture partners might use the power of the

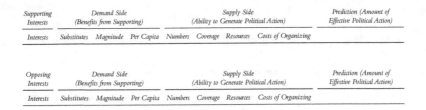

Supporting Interests	Demand Side *(Benefits from Supporting)*			Supply Side *(Ability to Generate Political Action)*				Prediction *(Amount of Effective Political Action)*
Interests	Substitutes	Magnitude	Per Capita	Numbers	Coverage	Resources	Costs of Organizing	

Opposing Interests	Demand Side *(Benefits from Supporting)*			Supply Side *(Ability to Generate Political Action)*				Prediction *(Amount of Effective Political Action)*
Interests	Substitutes	Magnitude	Per Capita	Numbers	Coverage	Resources	Costs of Organizing	

Figure 1.6   The Distributive Politics Spreadsheet

state against them. This work, then, combines market, firm, and nonmarket analysis in an interesting way.

For our purposes, focusing on contractual hazards provides us with some insights on how firms might organize both their trade and investment activities. Figure 1.5 suggests a possible array of organizational forms that should vary with asset specificity concerns, ceteris paribus nonmarket factors, and hazards.

As this chart suggests, for trade, organizational forms will vary from an arm's length trade where contractual hazards are few to trade wholly within a firm where there is great concern about such hazards. Similarly, for investment, contractual hazards could be mitigated by higher levels of ownership, albeit with the negative costs involved with maintaining a bureaucratically organized firm. Although our primary focus in this book is not on firm organization and structuring, these ideas should provide some insights into organizational responses to market and nonmarket factors.

Nonmarket Strategy

In order to systematically assess the supporting and opposing interests involved in a particular nonmarket issue, we can consider the following "distributive politics spreadsheet" in figure 1.6.[21] This approach divides the benefits and costs to each side from supporting or opposing a particular course of action on an issue that may have consequences for a firm.

Based on the well-known literature on interest group politics, this figure provides a mode of assessing the likely effective political action that can be generated by groups on each side of an issue. The demand side looks at the incentives for varying interest groups based on substitutes (alternatives to engaging in action on this issue), the overall magnitude of benefits arising from success in the issue, and the per capita benefits that specifically speak to the motivation of a particular interest group.

The supply side considers the power capabilities of the actors in question, focusing on their numbers (how many groups or individuals can be involved), the coverage in terms of relevant political jurisdictions, and the resources that can be brought to bear on the issues. The last element, the cost of organizing, speaks to the problems of overcoming collective action in view of the possibility of free-riding and information dissemination. Such an analysis can be conducted for both the supporting and opposing sides on any issue. The definitions of what issue or issue-areas are involved, as well as which groups or individuals should be considered relevant political actors, depend of course on the problem that is being addressed, and particularly on the geographical arena in question.

Integrated Strategic Choice

I have argued that firms must make strategic decisions about the arena in which to position themselves; but they must also decide where to locate within a particular arena. As an example, a firm needs to make a decision as to whether to concern itself with the cost/quality decision at the national or regional level. The success of this policy from a market perspective will depend on the likelihood of entrants, be they domestic or global. This factor could potentially be controlled through market actions and organizational strategies, thus moving the firm to a position on the cost/quality dimension to prepare the firm for severe competition, even from potential global competitors. Alternatively, firms may try to insulate a national level playing field or regional arena through nonmarket protectionist actions. The investment decision in market competitiveness versus investment in political strength is one that firms must make on an ongoing basis. To take another example, firms in the telecommunications industry, faced with deregulation and new competition, have tried to position themselves globally through setting standards (through the

Comité Consultatif International Téléphonique et Télégraphique [CCITT] in the International Telecommunication Union [ITU]) as well as engaging in buyouts in other countries, alliances, and the like. This involves positioning oneself for timing (standard setting) as well as in the cost/quality and strongholds arenas.

To get a better handle on integrated choice, we can briefly consider some examples of how firms combine market and nonmarket strategies. Faced with nonmarket problems such as Japanese banking regulations, European banks worked with their governments to pressure the Japanese to provide reciprocal access to their markets while introducing new types of products to enhance their competitiveness. Similarly, while European firms in the auto industry have attempted to design products that will be attractive to Asian markets, they have also increasingly turned to their home governments to lobby the Chinese in an effort to secure contracts and investment projects in this giant market.

At the broader regional level, the U.S. efforts in APEC and the discussions about a possible East Asian grouping have stimulated broad concern among European firms. In response, these firms helped to push the agenda for meetings between the European Union (EU) and Asian countries in the Asia-Europe Meeting (ASEM). This group came together in March 1996 (and has met every two years since then) with twenty-five countries in attendance: fifteen EU members, seven ASEAN members, South Korea, Japan, and China. Of particular note is an agreement to set up an "Asia-Europe Business Forum" to promote greater two-way trade and investment between Asia and Europe. European firms with strong competitive positions in key industries such as power generation equipment and telecommunications infrastructure have been especially interested in encouraging a dialogue in this area. Finally, we can view Airbus's efforts in Asian markets as both a market and nonmarket response to both market and nonmarket factors. The European consortium has made efforts to compete with Boeing by lobbying European home governments, the European Community, Asian governments, and even the U.S. government. At the same time, Airbus has attempted to enhance the appeal of its products through technological advancements, and specific interfirm alliances in the Asia-Pacific region.

To sum up: these examples provide an illustration of the types of issues that firms have faced and are likely to face increasingly in the Asian region. As they attempt to enhance their competitive position, they are likely to draw upon an array of specific market and nonmarket strategies. The lessons they learn from similar efforts by firms from other countries should prove instructive.

IV. Tactical Analysis:
Implementing Strategy

To implement a dynamic strategy successfully, firms must focus on three different tasks. The first is to undertake market strategies through the development and use of their capabilities. A second task involves executing nonmarket strategies, both as an adjunct to their market strategies as well as to create strategic competitive space for a longer-term market strategy. And finally, firms must utilize and restructure their organization to fit with their dynamic market and nonmarket strategies and to position themselves for new opportunities. The necessary tasks are depicted in figure 1.7.

Market Tactics

With respect to market strategy implementation, we can focus on three elements of firm tasks: research and development (R&D), production, and marketing. In positioning themselves in various arenas (say cost/quality and timing/know-how), firms must decide how best to compete. Thus, if a decision is pursued to compete with other multinationals using know-how, it is self-evident that emphasis might be placed on R&D. Thus, for example, Japanese firms have located their design centers for automobiles in the Los Angeles area to take advantage of the superior resources in that region and to market autos better in the U.S. market. To take another example, European firms who are focusing on China face the problem of marketing their products to compete with the many other contenders in that increasingly crowded market as firms from many countries attempt to tap into the large domestic market. Or, in choosing to use production networks across a number of Asian countries, European firms must decide where to conduct R&D and choose an appropriate market to lower their costs without excessively sacrificing quality.

Organizational Tactics

With respect to tactics, having chosen an appropriate form of trade or investment in light of transaction costs considerations, firms must internally restructure their management and firm organization to succeed in the market arena they have chosen. Wholly owned subsidiaries require knowledge of sourcing partners and personnel who understand local markets and who can deal with host governments. Some of these tasks could be shifted to a local partner in a joint venture. In such cases, however, skill in organizing

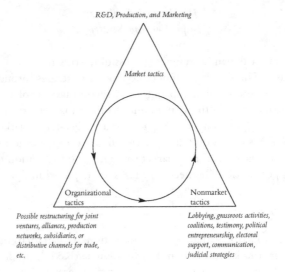

Figure 1.7   Tactical Anaylsis

and managing joint ventures with respect to contracting, financing, and control will be necessary. With respect to trade strategies, firms must again organize themselves to compete effectively. The failed effort by Sears to compete with Japanese trading companies illustrates the problems of operating in highly competitive markets and the need for organizational skill and learning.

With respect to nonmarket strategies and tactics, firms must develop skill in interacting with governments, nongovernmental organizations, and other interest groups. This aspect of developing an integrated strategy and implementing one successfully has often been neglected by firms who concentrate only on market issues and attempt to outsource nonmarket tasks.

Nonmarket Tasks

Eight implementation elements for dealing with nonmarket problems are lobbying, grassroots activity, coalition building, testimony, political entrepreneurship, electoral support, communication and public advocacy, and judicial strategies.[22] For the most part, these are self-explanatory. Some may be less obvious. Grassroots activities refer to efforts to generate broad scale public support, with the objective of influencing office holders. Another notion that might be more opaque is political entrepreneurship. This tactic involves the active role of attempting to set a political agenda of relevance to the firm. Examples include efforts to open markets in Japan, putting intellectual property issues on the GATT agenda in the Uruguay Round, and promoting arrangements such as NAFTA. In most cases, entrepreneurship of this type involved coalition building with like-minded firms as well as the host of other tactical efforts to secure success in this agenda setting process.

V. Layout of the Book

How have European firms fared in Asia? To examine this issue, contributors to this volume examine the strategies of European firms in different industries. To provide a context for this discussion, in Part II of the book, three authors provide a systematic economic and political analysis of the constraints and opportunities facing European firms.

In chapter 2, Shujiro Urata provides an analysis of recent trends in aggregate trade, FDI, and strategic alliances in Asia. Urata finds that European investors have not kept up with their U.S. and Japanese counterparts in terms of the growth in total stock of foreign investment in East Asia. European investors have focused on a few specific countries in Asia, with over 60 percent of European FDI going to the ASEAN-4 and over half of this amount ending up in Indonesia. To explain this uneven performance, Urata reevaluates the factors identified by a 1996 EU-UNCTAD report on impediments to investment in Asia including regulatory regimes, structural characteristics of the economy, transaction costs, government support, investment strategies, and preoccupation with regional integration. He concludes with a discussion of European efforts to overcome such barriers and prospects for increasing investment.

In chapter 3 John Ravenhill outlines the nonmarket context in Asia by focusing on the implications of Asian regional integration efforts for firm strategies. Ravenhill provides a brief summary of the recent evolution of the three major regional groupings in which a number of East Asian countries participate: ASEAN, APEC, and ASEM. In emphasizing the fledgling and weakly institutionalized nature of these groupings, Ravenhill explores the issue of the appropriate level at which firms might most successfully lobby for market access and benefits.

As a complement to Ravenhill's emphasis on the external Asian nonmarket environment, in chapter 4 Cédric Dupont focuses on the internal domestic lobbying environment within the EU. He devises a decision map to examine potential firm strategies for lobbying in the EU. In particular, Dupont focuses on the appropriate target for lobbying, the most profitable route, and the organizational form likely to yield the highest return. He then turns to an examination of European firms' experiences in a variety of sectors to provide an empirical complement to the analytical structure that he develops.

Part III of the book turns to an examination of European firm strategies in five sectors: software, autos, commercial aircraft, banking, and insurance (the final two are covered in a single chapter). Chapter 5, by Trevor H. Nakagawa, explores the performance of European software firms. He finds that European firms have been slow to enter Asian software markets, and in fact have been steadily losing market shares to U.S. firms in their own domestic markets. Nakagawa shows that despite significant aid from their governments, fragmentation and specialization along national, environmental, and functional divisions have hindered European software firms. After exploring the market and nonmarket factors that account for this relatively poor European performance, Nakagawa then considers the cases of three European firms that have been much more successful than the European norm: Systems, Applications, Products (SAP) in Enterprise Resource Planning software, Micro Focus in Common Business-Oriented Language (COBOL) development tools, and Synon in the development of Computer Aided Software Engineering (CASE) tools. In analyzing the success of these firms, he is able to provide insight into the combination of market and nonmarket strategies that have allowed them to succeed.

The automobile industry, for its part, has seen intense, global competition among European, Asian, and American firms. Among European producers, there have been sharp differences in success, thus providing a ready study of the market and nonmarket factors that are instrumental for auto firms to compete effectively. Chapter 6, by Beverly Crawford and Nick Biziouras, examines the efforts of Peugeot and Volkswagen (VW) in one of fastest growing markets in Asia—China. As they demonstrate, VW implemented a carefully formulated integrated market and nonmarket strategy and was able to establish a strong market position in China, while Peugeot eventually sold its production facilities. They argue that Peugeot's failure can be attributed to inadequate customization of product for local needs, inability to develop a strong distribution network, and failure to cooperate with local authorities. By contrast, VW succeeded with a strong distribution network, first-mover advantage, and the use of local management talent and local components suppliers. Therefore, the VW approach illustrates at least one winning combination of market and nonmarket strategies for success in the Chinese auto market.

Chapter 7, by William Love and Wayne Sandholtz, explores the efforts of Airbus to counter Boeing's success in Asia. As the fastest growing market in the world and facing no indigenous competitors, Asian markets were a key Airbus target. Indeed, Airbus did succeed in penetrating Asian markets using a strategy that emphasized both market and nonmarket factors. From a market perspective, Airbus sought to develop customer goodwill by lending aircraft and pilots to Asian airlines and focusing on efficient, high-capacity, technologically advanced planes to deal with the limited supply of landing spots at airports in most Asian countries. From a nonmarket perspective, Airbus relied on "launch aid" from national governments, entered into subcontracting relationships with governments to encourage purchase by their flagship airlines, and pushed European governments to offer technical and developmental assistance to governments whose airlines chose Airbus.

Chapter 8, by Klaus Wallner, examines European service sector firms in Asia. Specifically, he focuses on the strategies and performance of European banks in Southeast Asia and European casualty insurance providers in Japan. Wallner finds that European banks faced greater difficulty in deploying European governmental pressures on their behalf, in marked contrast to American and Japanese banks. However, he shows how, through a combination of direct market and nonmarket strategies, European banks were able to develop their relationship with Asian clients to raise their prospects of securing the corporate bond and investment banking business. By contrast, in the casualty insurance business, European firms were unable to compensate for their relative lack of success in securing government support by deploying market strategies in the face of a highly regulated Japanese insurance market.

The conclusion (Part IV) of the book assesses how European firms have penetrated Asian markets in the face of changing market and nonmarket conditions.

Notes

1. For comments on earlier versions of this chapter, I would like to thank Cédric Dupont, Kun-Chin Lin, Trevor Nakagawa, and Ralph Espach.
2. See Baron (1999) and (2000) for an excellent discussion of nonmarket strategies.

3. These categories are taken from Baron (2000).
4. Bartlett and Ghoshal (1989).
5. See chapter 3 by Ravenhill in this volume.
6. See, for example, Aggarwal (1992).
7. Porter (1980).
8. It is worth noting that other analysts have criticized Porter's approach for being excessively structural and unresponsive to firm strategies. This debate, similar to the "Great Man" vs. "Forces of History" argument in both political science and history, concerns the plasticity of structural forces as opposed to the initiative that firms might take to mold the factors themselves.
9. See Aggarwal, Keohane, and Yoffie (1987).
10. See Bain (1956).
11. Hill and Jones (1995).
12. Prahalad and Hamel (1990), p. 82.
13. Hamel and Prahalad (1989).
14. Baron (2000), pp. 11–15. The four I's noted here provide a useful but limited first cut to understand the nonmarket environment as I discuss in the following paragraphs.
15. See among others, Haas (1980), Stein (1980), Oye (1979) and Aggarwal (1996).
16. D'Aveni (1994).
17. D'Aveni (1994), p. 22.
18. See Coase (1960) and Williamson (1985, 1996), among others.
19. See the excellent work by Henisz (1998), who draws upon Oliver Williamson's work on economic transaction costs and work by Douglass North (1981, 1999) on political transaction costs to examine organizational form choices for direct foreign investment in the context of possible expropriation.
20. Klein and Leffler (1981) and Henisz (1998).
21. See Baron (2000) for discussion of the "distributive politics spreadsheet."
22. See Baron (2000) for a discussion.

References

Aggarwal, Vinod K. (1992). "The Political Economy of Service Sector Negotiations in the Uruguay Round." *The Fletcher Forum of World Affairs* 16 (1) (Winter), pp. 35–54.

———. (1996). *Debt Games: Strategic Interaction in International Debt Rescheduling.* New York: Cambridge University Press.

Aggarwal, Vinod K., Robert Keohane, and David Yoffie (1987). "The Dynamics of Negotiated Protectionism," *American Political Science Review* 81 (2) (June), pp. 345–366.

Aggarwal, Vinod K. and Charles Morrison, eds. (1998). *Asia-Pacific Crossroads: Regime Creation and the Future of APEC.* New York: St. Martin's Press.

Bain, J. E. (1956). *Barriers to New Competition*. Cambridge: Harvard University Press.

Baron, David (1999). "Integrated Market and Nonmarket Strategies in Cline and Interest Group Politics," *Business and Politics* 1 (1) (April), pp. 7–34.

Baron, David (2000). *Business and Its Environment,* 3rd edition. Upper Saddle River, NJ: Prentice Hall.

Bartlett, C. and S. Ghoshal (1989). *Managing Across Borders: The Transnational Solution*. Boston: Harvard University Press.

Coase, Ronald (1960). "The Problem of Social Cost." *Journal of Law and Economics* (October), pp. 186.

D'Aveni, R. (1994). *Hypercompetition: Managing the Dynamics of Strategic Maneuvering*. New York: The Free Press.

Haas, Ernst (1980). "Why Collaborate? Issue-Linkage and International Regimes." *World Politics* 32 (3), pp. 357–405.

Hamel, Gary and C. K. Prahalad (1989). "Strategic Intent." *Harvard Business Review* (May-June), pp. 63–76.

Hill, Charles and Gareth Jones (1995). *Strategic Management Theory: An Integrated Approach*. Boston: Houghton Mifflin.

Henisz, Witold (1998). "The Institutional Environment for Multinational Investment," paper presented at the Positive Political Theory Conference.

Klein, Benjamin and Keith B. Leffler (1981). "The Role of Market Forces in Assuring Contractual Performance." *Journal of Political Economy* 89 (4), pp. 615–641.

North, Douglass C. (1981). *Understanding the Process of Economic Change*. London: Institute of Economic Affairs.

North, Douglass C. (1999). *Structure and Change in Economic History,* 1st edition. New York: Norton.

Oye, Kenneth (1979). "The Domain of Choice." In *Eagle Entangled: U.S. Foreign Policy in a Complex World*. New York: Longman, pp. 3–33.

Oye, Kenneth, Robert Lieber, and Donald Rothschild, eds. (1979). *Eagle Entangled: U.S. Foreign Policy in a Complex World*. New York: Longman.

Prahalad, C. K. and Gary Hamel (1990). "The Core Competence of the Corporation." *Harvard Business Review* (May-June), pp. 79–91.

Porter, M. E. (1980). *Competitive Strategy*. New York: Free Press.

Stein, Arthur (1980). "The Politics of Linkage." *World Politics* 33 (1) (October).

Williamson, Oliver E. (1985). *The Economic Institutions of Capitalism: Firms, Markets and Relational Contracting*. New York: The Free Press.

Williamson, Oliver E. (1996). *The Mechanisms of Governance*. New York: Oxford University Press.

Part Two

The Asian and European Market and Nonmarket Environments

Chapter 2

Europe's Trade and Foreign Direct Investment in Asia

Shujiro Urata

I. Introduction

Until Asia was abruptly hit by a financial and economic crisis in 1997, it had been a center of phenomenal economic growth in the post–World War II period.[1] In the 1950s and 1960s Japan achieved remarkable economic growth, while in the 1970s and 1980s the Asian newly industrializing economies (NIEs)—consisting of Hong Kong, South Korea, Singapore, and Taiwan—experienced similarly impressive growth rates. In the mid-1980s several Association of Southeast Asian Nations (ASEAN) member countries (in particular Indonesia, Malaysia, and Thailand) joined the Asian NIEs on the path of rapid economic growth. In the 1990s, these fast-growing economies were joined by China and other Southeast Asian economies.

Several factors, including sound macroeconomic policies, highly educated populations, high savings rates, and outward-oriented development strategies, have been identified as catalysts for the rapid economic growth of Asian economies in the post–World War II period.[2] Among these factors, expansion of foreign trade and, more recently, foreign direct investment (FDI), have played a particularly important role in promoting economic growth. Through foreign trade, Asian economies have obtained capital goods, intermediate goods, technologies and other items necessary for economic development. FDI has brought recipient Asian economies not only funds for investment but also technologies and managerial know-how, enabling them to expand production capabilities and begin to realize their economic potential.

In July 1997, the Thai baht experienced a sudden, sharp depreciation and other Asian countries soon saw their currencies dragged down in its wake, sparking a financial and economic crisis throughout Asia.[3] Many Asian economies now appear to have recovered from the crisis, and since the fundamentals for economic growth remain in place, successful implementation of structural reforms to strengthen financial and corporate sectors should enable Asian economies to achieve once again high rates of economic growth.

In contrast to Asia, Europe experienced relatively subdued economic growth through the mid-1980s.[4] In the latter half of the 1980s, the European Community (EC) started to revitalize itself through removal of trade barriers and admission of new members. In the early 1990s, the European economy suffered a recession, but recovered by the mid-1990s. Since then, economic growth has been relatively steady despite a slight ripple caused by the 1997–98 currency crises in developing economies. Meanwhile, the establishment of the European Union (EU) in 1993 and the introduction of a common currency, the euro, in January 1999 have marked substantial steps toward achievement of Economic and Monetary Union (EMU). The EU continued to expand its membership in the 1990s, further increasing its weight on the world stage.[5] These recent developments toward European economic integration and expansion are substantially responsible for Europe's favorable economic performance in the last few years.

Mutual self-interest has helped to drive these two regions toward closer economic ties. Asian countries have been interested in promoting economic ties with Europe due to Europe's inherent attractiveness as an export market and its related value as a means to reduce Asian dependence on the American market. Europe is likewise interested in promoting economic ties with Asia, recognizing the benefits of a closer relationship with a fast growing region. Indeed, as Europe has made substantial progress toward economic integration, its interest in Asia has increased. However, Europe's presence in Asia in the form of foreign trade and FDI has been quite limited relative to the size of each region's economy; Japan and the United States have been much more important trade and FDI partners for Asia. Nevertheless, there is every indication that Europe and Asia will intensify transregional economic activities in the coming years.

In light of these efforts toward closer commercial ties, this chapter aims to examine several facets of the emerging Asia-Europe economic relationship. First, Sections II and III review the patterns of European trade and FDI in Asia in recent years to set the stage for the sectoral analysis in this volume. Second, Section IV attempts to discern the factors behind Europe's relatively low trade and investment presence in Asia as compared to the United States and Japan. Third, Section V discusses various policies and measures taken by Europe to expand its involvement in economic activities in Asia. Finally, the chapter concludes with an examination of future prospects for an expanded European economic presence in Asia.

II. Foreign Trade and Foreign Direct Investment in Asia

Until the financial crisis erupted in 1997, foreign trade and FDI in Asia had been expanding rapidly since the 1980s.[6] Including Japan, between 1980 and 1997 Asian trade (exports and imports) expanded approximately five-fold, as compared to world trade, which roughly tripled.[7] These trade figures become even more impressive when Japan is excluded, with Asian trade expanding by a multiple of 6.7 during the same period. Meanwhile, the stock of Asian FDI (outflows and inflows) grew by 13 times between 1980 and 1994, at a significantly greater rate than world FDI stock, which increased 5.7 times. Within Asia, including Japan, the stock of outward FDI expanded 15 times, while inward FDI grew tenfold.[8] As a result of rapid expansion of foreign trade and FDI by Asian economies, Asia's share in world trade and FDI increased significantly. Specifically, the shares of developing Asia in world exports and imports increased from 7.5 and 7.4 percent respectively in 1980 to 12.3 and 11.4 percent in 1997, and their shares in world outward and inward FDI stock increased from 1.7 and 7.6 percent respectively in 1980 to 6.3 and 14.2 percent in 1994.[9]

Various factors explain the rapid expansion of foreign trade and FDI in Asia during the mid-1980s to mid-1990s. One of the most important is a shift from an inward to an outward orientation of trade and FDI policies. In the environment of the Latin America debt crisis in the early 1980s, developing East Asian economies adopted structural adjustment policies primarily consisting of liberalization of foreign trade and FDI and supplemented by deregulation of domestic economic activities. The change from inward-looking protectionism to outward-looking liberalization was partly attributable to the recommendations by donors such as the World Bank and the International Monetary Fund (IMF), and partly to Asians' own recognition of the part that liberalization played in the success of the NIEs. Their actual implementation of market-oriented policies primed the Asian economies for the subsequent expansion of exports and FDI inflow. Trade liberalization shifted the incentives from import-substitution to export production, and FDI liberalization increased the attractiveness of these economies to foreign investors.

In addition to the purposive implementation of trade and FDI liberalization policies, several external developments in the mid-1980s facilitated the expansion of exports and FDI inflow in Asia. One is the substantial realignment of the exchange rates of major currencies. In September 1985, to correct the imbalances in their respective current accounts, the Group of Five industrial countries (G-5) agreed to realign their exchange rates to increase the value of the Japanese yen and Deutsche mark relative to the U.S. dollar. The yen's appreciation contributed to export expansion of Asian economies, increasing the prices of Japanese products vis-à-vis those from other Asian economies and thus stimulating Asian exports not only to Japan but to other markets as well.

The drastic yen appreciation stimulated Japanese FDI to Asian economies in two ways.[10] First, to cope with the loss in international competitiveness, many Japanese firms moved their production base from Japan to foreign economies where production costs were lower. Second, Japanese firms were experiencing a "liquidity" or "wealth" effect, wherein increased collateral and liquidity enabled them to finance FDI more cheaply relative to their foreign competitors. During the second half of the 1980, the Japanese government was injecting liquidity into a domestic economy suffering from a recession caused by declining exports, thereby inflating the prices of shares and land and initiating the so-called "bubble economy." This increase in liquidity and the subsequent asset-price inflation further promoted Japan's FDI in Asia by making it easier for Japanese firms to obtain loans and finance investments abroad. It also contributed to the expansion of exports from Asian economies to Japan by increasing Japanese demand for imports.

An appreciating yen did not end Japan's trade friction with the United States and the European Community, however, making FDI more attractive than trade for Japanese firms. In order to secure markets in developed economies, a number of Japanese firms invested not only in Europe and the United States but also in other economies, most notably in Asia. This strategy allowed them to set up export platforms and thus to get around the import barriers in developed economies.[11]

While Japanese firms contributed to the rapid expansion of exports and FDI in Asia during the second half of the 1980s, the newly industrializing economics followed Japan's footsteps in contributing to the rapid expansion of intraregional exports and FDI in the 1990s. With their own national economies facing a similar set of problems such as currency appreciation, speculative bubbles, and trade friction with developed economies, firms from the NIEs also invested in the ASEAN countries, China, and other parts of Asia in search of low-cost production.

In sum, a notable cause of rapid economic growth in Asia in the pre-crisis period was the formation of a trade-FDI nexus. Trade and FDI expansion interacted to contribute to growth by increasing and improving productive capability and by expanding demand for manufactured products. These factors helped Asian economies experience a benevolent circle in which expanding trade, investment, and growth reinforced one another. Meanwhile, economic growth at home enabled Asian countries not only to promote outward FDI and exports by expanding technical, financial, and managerial capabilities, but also to absorb inward FDI and imports by providing increasingly attractive domestic markets.

Although the economic crisis of 1997–98 had serious ramifications for the Asian economies in the form of reduced output and employment, its impact on FDI was not substantial. Indeed, compared to portfolio investment and bank loans, which declined precipitously, the decline in FDI inflows was small, indicating the resilience of FDI among different types of financial flows.[12] In particular, non-Asian investors became increasingly important suppliers of FDI, as the crisis temporarily diminished intra-Asian FDI.

In trade, substantial currency depreciation stimulated export expansion and import contraction, contributing to an improvement in the trade balances of many East Asian economies. Export expansion played a pivotal role in reducing the negative impact of the crisis for many of those affected, allowing Asian economies to increase exports to extra-regional countries and thus to lessen the impact of the contraction of intraregional demand.

III. European Firms in Asia

This section considers the extent to which European firms participated in Asia's economic expansion. More specifically, it examines the European position in foreign trade and FDI in Asia. As we shall see, Europe has played a relatively smaller role in Asia than might be expected, and thus the next section attempts to identify the factors leading to this underrepresentation.

Foreign Trade

Europe's overall foreign trade has expanded steadily in recent years. Between 1980 and 1998, European exports and imports (in terms of U.S. dollars) increased by 3 and 2.6 times, respectively, roughly in line with a global trade expansion of 2.9 times (see table 2.1).

Particularly notable is the large share of intra-European trade—approximately 60 percent of it is conducted within the region. Although European trade with Asia has increased at a faster rate than with other regions, it accounts for a very small portion of total European trade. In 1998 as little as 4.3 percent of European exports were destined for Asia, while Asia accounted for 6.8 percent of European imports—indicating that Asia has played a more important role as an import source for Europe than as an export destination.

Though the statistics examined above show the status of East Asia as a trading partner for Europe, they do not provide any information concerning the relative importance of Asia for Europe. To examine this, the "gravity coefficients" are computed.[13] The estimated gravity coefficients for bilateral trade between Europe and Asia during the 1970–97 period ranged from 0.2 to 0.4, indicating that their trading relationship is weak relative to overall world trade.[14]

A typology of Europe's exports to Asia reveals a relatively high concentration of manufactures compared to its exports to the rest of the world (see table 2.2). In 1996, 89 percent of Europe's exports to Asia were manufactures, while the corresponding value for its overall exports was 78 percent. Of these, machinery and transport equipment accounted for half of Europe's total exports to Asia.[15]

Accordingly, Asia is not a primary trading region for Europe. When the roles are reversed, however, the outcome is quite different. Table 2.3 shows that in 1998 Asia's exports to Europe accounted for 15.3 percent of its overall exports, while Asia's imports from Europe constituted 12.8 percent of its overall imports.[16]

From these observations one can see that Europe is a more important trading partner for Asia than vice versa. However, the importance of Europe in Asia's trade declined in the 1980s and 1990s, mainly due to a rapid increase in intraregional trade in Asia during the period.

Foreign Direct Investment

In the eight years from 1985 to 1993, overall global FDI more than tripled (see table 2.4). Various factors contributed to this global increase

Table 2.1 EU Trade with Asia

Value ($ millions)	Exports ($ millions)			Imports ($ millions)		
	1980	1990	1998	1980	1990	1998
Asia	17,259	53,908	96,168	22,340	65,397	141,223
NIEs	8,896	32,739	57,138	13,319	44,756	76,600
Hong Kong	3,147	8,861	20,268	4,869	15,232	26,372
Korea	1,439	8,269	9,549	2,931	10,028	18,213
Singapore	2,741	7,571	12,100	2,558	7,935	17,404
Taiwan	1,569	8,038	15,221	2,961	11,561	14,611
ASEAN4	5,637	13,796	19,651	6,531	14,362	36,461
Indonesia	1,894	3,831	4,319	1,441	3,094	8,907
Malaysia	1,653	3,491	5,993	2,325	4,525	11,885
Philippines	926	1,652	3,516	1,039	1,516	5,995
Thailand	1,164	4,822	5,823	1,726	5,227	9,674
China	2,726	7,373	19,380	2,491	6,279	28,162
Japan	7,283	30,911	35,029	19,804	58,701	71,720
NAFTA	50,015	122,923	203,645	72,811	117,632	164,021
U.S.	40,906	104,427	176,905	61,594	103,488	149,813
EU	459,470	985,137	1,347,290	459,470	985,137	1,347,290
World	753,772	1,492,233	2,232,600	807,782	1,480,377	2,073,100

Share (%)	Exports ($ millions)			Imports ($ millions)		
	1980	1990	1998	1980	1990	1998
Asia	2.29	3.61	4.31	2.77	4.42	6.81
NIEs	1.18	2.19	2.56	1.65	3.02	3.69
Hong Kong	0.42	0.59	0.91	0.60	1.03	1.27
Korea	0.19	0.55	0.43	0.36	0.68	0.88
Singapore	0.36	0.51	0.54	0.32	0.54	0.84
Taiwan	0.21	0.54	0.68	0.37	0.78	0.70
ASEAN4	0.75	0.92	0.88	0.81	0.97	1.76
Indonesia	0.25	0.26	0.19	0.18	0.21	0.43
Malaysia	0.22	0.23	0.27	0.29	0.31	0.57
Philippines	0.12	0.11	0.16	0.13	0.10	0.29
Thailand	0.15	0.32	0.26	0.21	0.35	0.47
China	0.36	0.49	0.87	0.31	0.42	1.36
Japan	0.97	2.07	1.57	2.45	3.97	3.46
NAFTA	6.64	8.24	9.12	9.01	7.95	7.91
U.S.	5.43	7.00	7.92	7.63	6.99	7.23
EU	60.96	66.02	60.35	56.88	66.55	64.99
World	100.00	100.00	100.00	100.00	100.00	100.00

Note: Asia includes the NIEs, ASEAN4, and China.

Source: Computed from trade data compiled by Japan Trade Organization

Table 2.2 Europe's Exports to Asia by Products: 1996

	Export Value ($ billions)		Product Composition (%)		
	Asia (1)	World (2)	Asia	World	(1)/(2)
Agricultural products	14.95	246.29	6.8	10.8	0.061
Food	12.18	209.25	5.6	9.2	0.058
Raw materials	2.77	37.04	1.3	1.6	0.075
Mining products	5.48	144.39	2.5	6.3	0.038
Ores and other minerals	1.44	15.83	0.7	0.7	0.091
Fuels	0.98	90.61	0.4	4.0	0.011
Non-ferrous metals	3.06	37.95	1.4	1.7	0.081
Manufactures	194.49	1774.77	89.0	77.8	0.110
Iron and steel	5.73	70.88	2.6	3.1	0.081
Chemicals	27.9	276.9	12.8	12.1	0.101
Other semi-manufactures	20.25	214.51	9.3	9.4	0.094
Machinery and transport equipment	108.03	869.26	49.5	38.1	0.124
Automotive products	17.35	246.08	7.9	10.8	0.071
Office and telecom equipment	21.8	175.52	10.0	7.7	0.124
Other machinery and transport equipment	68.89	447.66	31.5	19.6	0.154
Textiles	4.62	65.35	2.1	2.9	0.071
Clothing	4.57	59.87	2.1	2.6	0.076
Other consumer goods	23.38	217.99	10.7	9.6	0.107
Total merchandise exports	218.46	2281.80	100.0	100.0	0.096

Note: Asia in this table includes all the countries in Asia, including Japan. Total merchandise exports includes unspecified products.
Source: WTO, *Annual Report* 1999

in foreign investment. Technological progress in telecommunications services lowered the cost of communications, enabling multinational firms to extend their business networks by establishing overseas subsidiaries. Equally important, liberalization of FDI policies and privatization of public enterprises provided multinationals with greater FDI opportunities.

The foreign investment of European firms has reflected these global trends. The stock of European FDI in Asia rose $7.7 billion in 1985 to $21 billion in 1993, while its overall FDI tripled.[17] Since the rate of increase in European investment in Asia was lower than the corresponding rate for

Table 2.3 Importance of EU in Asia's Trade: Share of EU in Asia's Overall Trade (%)

	Exports			Imports		
	1980	1990	1998	1980	1990	1998
Asia	15.77	15.71	15.32	12.24	13.95	12.80
NIEs	17.44	16.75	14.55	10.73	13.05	12.98
Hong Kong	24.69	18.54	15.18	14.18	10.88	12.77
Korea	16.81	15.42	13.77	7.32	13.98	11.78
Singapore	13.20	15.04	15.84	10.42	13.61	12.64
Taiwan	14.94	17.20	13.21	10.65	14.69	14.54
ASEAN4	13.85	16.63	17.25	14.37	15.95	12.58
Indonesia	6.58	12.05	16.50	16.95	21.58	19.44
Malaysia	17.94	15.38	16.18	15.15	13.76	10.24
Philippines	17.96	18.50	20.32	11.00	12.49	10.20
Thailand	26.54	22.66	17.75	13.33	15.99	14.21
China	13.73	9.99	15.33	14.39	15.03	12.52

Note: Asia includes the NIEs, ASEAN4, and China.
Source: Computed from trade data compiled by Japan Trade Organization.

overall European FDI, Asia's share of overall European FDI declined slightly from 4.4 percent in 1985 to 3.8 percent in 1993. This decline mainly reflects active intra-European investment, as the share of inward investment as a proportion of total European FDI increased from 30 percent in 1980 to 47 percent in 1994.[18] The estimated gravity coefficients for the FDI linkage between Asia and Europe were around 0.3 between 1980 and 1995, indicating that their FDI relationship is significantly lower than its expected level.

European firms' evolving position in Asia needs to be considered in the context of investment trends in Asia more generally. While the European share of total FDI inflow into Asia declined from 16.4 percent in 1980 to 12.9 percent in 1993 (see table 2.5), Japanese and U.S. FDI in the region experienced similar downward trends: their respective shares in total FDI in Asia declined from 25.1 and 16.0 percent in 1980 to 21.0 and 14.1 percent in 1993—though both of their shares still eclipse that of Europe.

Meanwhile, intra-Asian investment has expanded significantly since the 1980s—increasing from 42.5 percent in 1980 to 51.9 percent in 1993. This trend is particularly visible in China, as Asian investors accounted for 80

Table 2.4 Foreign Direct Investment by the Triad in Developing Asia ($ millions and percentages)

| | Stocks | | | | | | Flows | | | |
| | 1980 | | 1985 | | 1993 | | 1985–87 | | 1990–93 | |
	Value	Share	Value	Share	Value	Share	Value	Share	Value	Share
NIEs										
EU	1,638	20.6	2,839	16.7	9,786	18.2	231	11.7	877	17.7
Japan	2,098	26.3	4,993	29.4	17,477	32.5	777	39.2	1,300	26.2
United States	2,445	30.7	6,233	36.7	16,300	30.3	695	35.1	1,332	26.9
Triad total	6,180	77.5	14,065	82.9	43,563	81.1	1,702	85.9	3,509	70.8
All countries	7,971	100.0	16,967	100.0	53,738	100.0	1,982	100.0	4,956	100.0
ASEAN4										
EU	2,841	15.0	5,635	19.7	18,042	15.0	336	19.7	2,324	13.8
Japan	5,087	26.9	7,595	26.5	26,842	22.4	535	31.5	3,234	19.2
United States	1,840	9.7	3,760	13.1	11,637	9.7	293	17.2	1,524	9.0
Triad total	9,768	51.6	16,991	59.3	56,520	47.1	1,164	68.4	7,081	41.9
All countries	18,942	100.0	28,663	100.0	120,023	100.0	1,700	100.0	16,886	100.0
China										
EU	300	13.6	584	8.3	2,018	3.5	113	5.5	300	2.6
Japan	128	5.8	502	7.2	4,288	7.5	245	12.0	782	6.7
United States	372	16.9	1,106	15.8	4,680	8.2	312	15.2	830	7.1
Triad total	800	36.3	2,192	31.2	10,986	19.2	670	32.7	1,911	16.4
All countries	2,202	100.0	7,015	100.0	57,172	100.0	2,048	100.0	11,631	100.0

(continues)

Table 2.4 *(continued)*

	Stocks						Flows			
	1980		1985		1993		1985–87		1990–93	
	Value	*Share*	*Value*	*Share*	*Value*	*Share*	*Value*	*Share*	*Value*	*Share*
Total Asia										
EU	4,779	16.4	9,058	17.2	29,846	12.9	679	11.9	3,501	10.5
Japan	7,313	25.1	13,090	24.9	48,607	21.0	1,558	27.2	5,316	15.9
United States	4,657	16.0	11,099	21.1	32,617	14.1	1,299	22.7	3,686	11.0
Triad total	16,748	57.5	33,248	63.2	111,070	48.1	3,536	61.7	12,502	37.4
All countries	29,115	100.0	52,645	100.0	230,933	100.0	5,731	100.0	33,473	100.0

Note: Singapore is included in NIEs. ASEAN4 includes Indonesia, Malaysia, Philipines, and Thailand.
Source: European Commission and UNCTAD (1996), Table 1.7.

Table 2.5 FDI Stock of the World and the EU in Selected Developing East and Southeast Asian Economies, by Industry, 1993 ($ millions and percentages)

	Indonesia			Malaysia			South Korea			Total		
	World	Share of EU	EU	World	Share of EU	EU	World	Share of EU	EU	World	Share of EU	EU
Primary sector	8,769	567	6.5	675		0.0	42	7	16.7	9,536	578	6.1
Agriculture	2,217	384	17.3	—	—	—	24	1	4.2	2,278	387	17.0
Mining and quarrying	6,552	183	2.8	675		0.0	18	6	33.3	7,258	191	2.6
Secondary sector	40,897	4,805	11.7	30,409	5,699	18.7	7,796	1,783	22.9	91,236	13,703	15.0
Food, beverages, and tobacco	2,009	717	35.7	1,104	226	20.5	490	27	5.5	4,363	1,022	23.4
Textiles, leather, and clothing	4,783	73	1.5	1,492	38	2.5	318	24	7.5	7,033	166	2.4
Paper	5,217	432	8.3	1,255	228	18.2	169	6	3.6	6,856	675	9.8
Chemicals	14,168	2,432	17.2	4,250	1,328	31.2	2,428	676	27.8	23,476	4,999	21.3
Coal and petroleum products	—	—	—	6,492	2,610	40.2	683	526	77.0	7,197	3,148	43.7
Rubber products	—	—	—	744	190	25.5	—	—	—	1,233	242	19.6
Non-metallic mineral products	2,973	416	14.0	1,519	147	9.7	206	99	48.1	5,304	730	13.8
Metals	11,391	729	6.4	5,619	270	4.8	227	16	7.0	18,430	1,125	6.1
Mechanical equipment	—	—	—	873	214	24.5	728	190	26.1	2,968	469	15.8
Electrical equipment	—	—	—	5,344	357	6.7	1,503	110	7.3	11,182	907	8.1

(continues)

Table 2.5 (continued)

	Indonesia			Malaysia			South Korea			Total		
	World	Share of EU	EU	World	Share of EU	EU	World	Share of EU	EU	World	Share of EU	EU
Motor vehicles	—	—	—	611	32	5.2	918	104	11.3	1,595	150	9.4
Other manufacturing	356	7	2.0	1,107	59	5.3	127	5	3.9	1,599	70	4.4
Tertiary sector	17,959	2,335	13.0	—	—	—	4,687	522	11.1	28,246	3,327	11.8
Construction	2,618	517	19.7	—	—	—	65	2	3.1	2,872	526	18.3
Distributive trade	922	10	1.1	—	—	—	595	214	36.0	2,901	322	11.1
Transport and storage	1,602	526	32.8	—	—	—	59	1	1.7	2,106	527	25.0
Finance and insurance	—	—	—	—	—	—	1,179	209	17.7	2,443	461	18.9
Real estate	1,903	109	5.7	—	—	—	—	—	—	1,903	109	5.7
Other services	10,913	1,174	10.8	—	—	—	2,765	96	3.5	15,997	1,381	8.6
All industries	67,625	7,705	11.4	31,084	5,699	18.3	12,525	2,313	18.5	129,018	17,606	13.6

Source: European Commission and UNCTAD (1996), Table 1.8.

Table 2.6 European FDI Stock in the World and Developing Asia by Industry ($ millions and percentages)

	1985			1993		
	World	Developing Asia	Share of Developing Asia	World	Developing Asia	Share of Developing Asia
Primary sector	39,529	2,383	6.0	57,074	4,045	7.1
Agriculture	905	345	38.1	1,728	277	16.0
Mining and quarrying	2,376	3	0.1	4,632	44	0.9
Petroleum	36,249	2,035	5.6	50,714	3,725	7.3
Secondary sector	57,169	1,953	3.4	192,341	7,041	3.7
Food, beverages, and tobacco	6,279	413	6.6	27,493	668	2.4
Textiles, leather, and clothing	113	1	0.9	1,329	20	1.5
Paper	1,275	51	4.0	11,237	241	2.1
Chemicals	19,849	842	4.2	59,437	4,395	7.4
Coal and petroleum products	12	—	—	356	1	0.3
Rubber products	19	—	—	925	–13	–1.4
Non-metallic mineral products	49	—	—	4,332	4	0.1
Metals	3,368	74	2.2	7,980	69	0.9
Mechanical equipment	4,982	62	1.2	11,340	171	1.5
Electrical equipment	8,011	250	3.1	31,226	942	3.0
Motor vehicles	4,017	32	0.8	14,223	67	0.5
Other transport equipment	842	99	11.8	2,295	45	2.0
Other manufacturing	8,355	131	1.6	20,167	431	2.1

(continues)

Table 2.6 (continued)

	1985			1993		
	World	Developing Asia	Share of Developing Asia	World	Developing Asia	Share of Developing Asia
Tertiary sector	77,028	3,347	4.3	304,405	8,522	2.8
Construction	1,524	131	8.6	5,026	-12	-0.2
Distributive trade	26,523	735	2.8	58,755	1,899	3.2
Transport and storage	2,240	703	31.4	10,186	90	0.9
Communication	9	—	—	227	—	—
Finance and insurance	15,802	976	6.2	114,770	2,747	2.4
Real estate	473	—	—	2,485	3	0.1
Other services	30,458	801	2.6	112,956	3,796	3.4
All industries	173,727	7,683	4.4	553,820	21,025	3.8

Source: European Commission and UNCTAD (1996), Table 1.2.

percent of China's FDI stock in 1993.[19] The rapid increase of intra-Asian investment in recent years reflects improvements not only in the financial position of Asian firms but also in their managerial capability, both of which benefited from the region's rapid economic growth.

European firms' investment activities in specific countries have varied somewhat across the region. Among the Asian economies, the ASEAN 4 (Indonesia, Malaysia, Thailand, and the Philippines) has been the most attractive group. These countries received 60 percent of European FDI in Asia prior to 1993, with Indonesia being the largest recipient with more than half of European FDI in the region.[20] Among the NIEs, Singapore has been the largest recipient of European FDI, mainly because of its close historic ties with European economies and its investment-friendly policies.[21] Also, European FDI in South Korea has seen a significant increase in recent years—an increase at least partly attributable to the programs implemented jointly by South Korea and EU members to promote European investment.[22] However, Europe has been much less active in investing in China: its share of total investment in China declined from 13.6 percent in 1980 to 3.5 percent in 1993.

The presence of European FDI in Asia also varies significantly across the primary, secondary, and tertiary sectors. The European presence is strongest in the secondary sector (see table 2.6). In 1993, European firms' share of Asian FDI in that sector was 15.0 percent, slightly higher than their corresponding share overall of 13.6 percent.

Among the secondary subsectors, Europe invested more heavily in coal and petroleum products, chemicals, rubber products, and food, beverages, and tobacco than did other investors. It is also worth noting that Europe registered a relatively strong FDI position in agriculture, not surprisingly given its corresponding presence in the food sector. Among the tertiary subsectors, European firms have been active in the transport and storage, finance and insurance, and construction sectors.

Although up-to-date statistics on the pattern of foreign investment in Asia are not readily available, it seems clear that European FDI in the region started to rise during the mid-1990s. Among the large-scale European projects were a $4 billion investment by Germany's BASF (chemicals) in China, and a $2.1 billion investment by France's GEC Alsthom (transport equipment) in South Korea.[23] In addition to the obvious allure of Asia's growing markets for European firms, a range of EU programs such as the Asia-Invest Program and the European Community Investment Partners probably contributed to an increase in European interest in East Asia.[24] The Asian financial and economic crisis also provided opportunities

for European firms to penetrate Asian markets through FDI, as depreciation of Asian currencies reduced the costs of investment for European
firms.[25] At the same time, the devastating effects on Asian (and Japanese)
firms and banks decimated intra-Asian FDI and cleared away local competitors to European firms.[26]

In contrast to the relatively low level of European FDI in Asia prior to
the financial crisis, European banks were active in providing loans in Asia
during the period before 1997. Active lending by European (and Japanese)
banks in Asia was in response to rising demand for borrowing by Asian
firms, which were investing strongly with an expectation that good times
would last.[27] By June 1997, European banks' share of total private loans
provided to Asian economies was 40 percent, higher than the corresponding shares for Japanese (32 percent) or U.S. banks (8 percent).

IV. Obstacles to Trade and FDI in Asia for European Firms

In spite of recent increases, European firms still have weaker trade and investment positions in East Asia compared to their Japanese and U.S. counterparts. To better understand why, we must consider the obstacles to
European trade and investment in Asia.

Obstacles to Trade

Various factors explain the considerably limited trade relationship between
Europe and Asia. Geographical distance is one effective barrier to foreign
trade, as the cost of shipping products increases with the distance to their
destination. Indeed, Masahiro Kawai and Shujiro Urata found this to be a
factor in their empirical investigation of the patterns in Japanese trade. Europe's distance from East Asia makes shipping costly.[28] By contrast, trade
within Europe and Asia incurs much lower shipping costs, subsequently facilitating closer trading ties within the respective regions.

Second, Dale Boisso and Michael Ferrantino found that in general "cultural distance" (for them proxied by language) negatively affects bilateral
trading relationship.[29] Cultural distance—in addition to the historical relationship between Europe and Asia—may be relevant in explaining their relatively limited bilateral trade relationship. One example of cultural
differences includes the non-transparency of Asian business practices, which
are based on personal connections rather than contracts. Such distinctions
have made it more difficult for non-Asian firms to be successful in Asia.

Third, in the past many observers have pointed to high import tariffs and quotas as barriers to trade in Asia. However, these points are no longer warranted since most Asian economies have been liberalizing their trade regimes. The impetus for this liberalization has come from several sources: recommendations and/or conditions attached to the provision of economic assistance from international organizations such as the World Bank and the IMF, as well as international and regional trade agreements under the auspices of the General Agreement on Tariffs and Trade (GATT) and World Trade Organization (WTO) (multilateral) and the Asia-Pacific Economic Cooperation (APEC) (transregional) frameworks.[30]

Only in one exceptional case do outsiders experience trade discrimination in Asia. Under the ASEAN Free Trade Area (AFTA)—the only regional trading arrangement in East Asia—ASEAN members agreed to complete tariff and nontariff liberalization vis-à-vis other members by 2002, and phase out temporary exclusion lists from tariff and nontariff liberalization within five years. Although AFTA will have a discriminatory impact on nonmembers, it does not seem likely to have a significant impact on European trade in Asia, as intra-ASEAN trade accounts for only 4 percent of intra-Asian trade in 1998. Furthermore, the discriminatory impact is likely to decline in the future, since AFTA members are scheduled to reduce tariff and nontariff barriers vis-à-vis non–AFTA members with a recognition on the part of the members that intra–AFTA trade is a small portion of their total trade.[31]

Although the causality between foreign direct investment and foreign trade is unclear, bilateral FDI and foreign trade flows have been shown to be closely related.[32] As such, one could argue that the small magnitude of foreign direct investment in Asia by European firms seems to have contributed to the relatively limited magnitude of bilateral trade flows.

Obstacles to Investment

A 1996 EU-United Nations Conference on Trade and Development (UNCTAD) joint report examined six possible factors limiting European FDI in Asia in recent years: (1) the regulatory regime for FDI in Asia; (2) structural characteristics of Asian host countries; (3) transaction costs; (4) government support; (5) investment strategies; and (6) preoccupation with regional integration. Its main conclusion is that European firms' underestimation of the growth potential of Asian economies led them to underinvest in the region. The report further argues that a preoccupation with regional integration in Europe also discouraged its FDI in Asia. Which of these conclusions is most convincing?

Asian regulations on capital inflows have often been cited as a factor limiting FDI. Indeed, until the early 1980s, a number of Asian economies observed strict regulations on FDI inflows to prevent foreign firms from driving uncompetitive local firms out of the market. Yet most Asian economies shifted to pro-FDI policies in the mid-1980s for reasons previously noted: the effects of the Latin America debt crisis, the recommendation and conditions from international organizations such as the World Bank and the IMF, and the promise of economic growth and its positive contribution to financial resources, managerial expertise, and technology. Although some sensitive sectors have been exempted from FDI liberalization, most Asian economies have liberalized their investment regimes to the extent that they are not effective deterrents to European FDI in Asia.[33]

The 1996 EU report argues that certain "structural characteristics"—namely the firm-specific, highly complex, and non-uniform regulations that are subject to frequent changes—have put European firms at a disadvantage vis-à-vis Japanese firms. While there is some truth to this claim, its effects are not entirely one-sided. Japanese firms are more familiar with the policymaking style in Asia, but a large number of Japanese firms have experienced as much difficulty dealing with the complex, non-transparent, and capricious nature of regulations in Asia as their European counterparts. In this regard, overseas Chinese firms are in a better position, as they have already developed extensive overseas Chinese networks in Asia.

Transaction costs, which are influenced by such factors as cultural barriers and variations in economic systems, may also play a role in determining bilateral or regional FDI flows. Similar cultural background and economic systems in Asia are often argued to give an edge to Asian (including Japanese) firms over non-Asian (including European) firms investing in Asia. However, a lack of quantifiable indicators of these variables precludes precise testing of their validity. Furthermore, the assumptions underlying the transaction cost argument can be questioned. In contrast to what is often assumed by some western analysts, Asia is a heterogeneous region with distinctly different religious and cultural traditions. Similarly, there is not one but several different models along which the Asian economies are organized, further undermining arguments of a pan-Asian exceptionalism.

Another possible explanation for limited European FDI in Asia is a lack of support by European governments through such means as the provision of information and financial resources. This argument may have some salience, given that the United States and Japan actively provide government support to their national firms undertaking FDI in Asia. However, one should not overemphasize the impact of government support. First, it is mainly a private sector initiative or request that leads to the provision of government support for FDI, rather than the reverse. Governments themselves do not initiate the expansion of FDI. Second, Japanese government support is provided to Japanese firms undertaking FDI in many parts of the world, not just to those investing in Asia. While it is often argued that Japan's Official Development Assistance (ODA) is concentrated in Asia, and therefore acts as a catalyst to Japanese firms investing in Asia, the validity of this argument is questionable. Japanese ODA has targeted infrastructure projects such as transportation facilities that improve the overall environment for FDI inflows regardless of their source. Still, provision of information by the Japanese government concerning prospective FDI host economies and their rules on FDI has helped Japanese firms, particularly in the case of small and medium-sized enterprises. The U.S. government has also provided such information to American firms. European governments have generally not followed suit, which may be partially responsible for low European FDI in Asia.[34]

Finally, regional arrangements such as APEC, which are endorsed by member governments, seem to encourage FDI among APEC member economies. However, APEC has not discriminated against nonmember investment, and the possibility that any such effect will be seen in the future is doubtful for several reasons. First, APEC advocates open regionalism under which no discriminatory measures will be applied against nonmembers. Second, so far APEC has only established nonbinding investment principles, and the FDI-promoting effect appears to be limited. Indeed, there are some principles that are more restrictive than the WTO rules on FDI or trade-related investment measures, casting doubt on the magnitude of the FDI-promoting effect of APEC among its members.[35] As with trade liberalization, FDI measures endorsed by APEC members are not discriminatory against nonmembers and thus should not discourage FDI from Europe.

More than the previous enumerated factors, a lack of well-formulated FDI strategies on the part of European firms may best explain their limited FDI in Asia. A brief comparison with Japanese firms helps to illustrate the point. Large Japanese multinationals have carried out FDI strategies involving their group companies and subcontractors that have enabled them to construct efficient procurement and sales networks in Asia. The agglomeration of Japanese firms in Asia has attracted additional Japanese FDI and provided business opportunities to new Japanese investors.[36] In contrast, European FDI has not reached the critical mass at which the agglomeration effect is likely to be realized.

The preoccupation of European firms with EU integration appears to be another important factor in explaining low levels of European FDI in East Asia. The widening and deepening of the European Union has given European firms ample business opportunities in their own backyard. Moreover, the EU has provided subsidies to make relatively underdeveloped areas within the EU more attractive for investment.[37] With limited financial and human resources for undertaking FDI, EU firms have good reason to concentrate their investment within the EU and other European regions. Now that the European economy has reached a very high level of integration as a result of the establishment of a common currency, and Asian economies have recovered from the economic crisis, European firms are likely to put a greater emphasis on Asia.

V. European Efforts to Promote Economic Ties with Asia

In light of their underrepresentation in Asia, Europeans have contemplated and implemented a number of measures to promote involvement in Asia. Until the mid-1990s, individual European countries defined their own national strategies. However, after the ratification of the Maastricht Treaty, the European Union began to increase its efforts to carry out promotion measures in a unified way. The shift in policy to a more concerted approach was a direct result of deeper integration in the European Union. This more consolidated approach improved the effectiveness of specific FDI-promotion measures.[38] One may divide these measures into two categories: one type promotes economic ties in general, and the other focuses specifically on trade and FDI.

Of the general measures to promote economic ties between Europe and Asia, among the most notable are the EU's efforts to establish closer cooperation with Asian countries in training and educational programs in management and technology. For example, the European Union-ASEAN Junior European Managers Program was established to give young Asian and European managers work experience in the other region. The EU's emphasis on training and educational programs stems from its realization that the creation of human networks is of the utmost importance for the promotion of close economic relations. These training and educational programs have been received favorably by Asian countries, which understand the need to improve human resources to achieve sustainable economic growth.

However, the most important general measure for the promotion of economic ties between the two regions has been the establishment of the Asia-Europe Meeting (ASEM). ASEM gathers the heads of states from European and Asian countries to discuss issues of common concerns with an eye toward promoting their economic relationship. The first ASEM was held in Bangkok in 1996 to promote trade and FDI to strengthen their overall economic ties. [39] A second ASEM was held in London in 1998, and the third took take place in Seoul in October 2000. ASEM has been successful in increasing awareness on the part of business as well as the general public regarding Europe-Asia relations, and contributed to closer economic and cultural ties between the two regions.

Turning to more specific measures, one finds that European countries have recently introduced a number of innovative plans to promote European trade and FDI in Asia. In addition to increasing the number of missions sent to identify possible trade and FDI opportunities, the Europeans have established several new programs to expand transregional economic ties. The program that has received the most attention has been the Asia-Invest Program, established in 1996. Because a lack of information was recognized as a serious bottleneck for increasing business activities in Asia, this program has attempted to break this blockage by disseminating data on trade and FDI opportunities. The program also provides financing to support activities such as exploring business contacts and solving problems of doing business in Asia. One of the notable characteristics of the Asia-Invest Program is the active participation of the private sector, which has been important for providing services that prove helpful to the users.

These programs complement the previously established European Community Investment Partners (ECIP) framework, set up in 1988, which provides various types of support to promote FDI by small- and medium-sized enterprises (SMEs), including identification of potential joint-venture partners and feasibility-study loans. ECIP appears to be quite successful, with more than fifteen hundred actions undertaken between 1988 and 1996. [40] Also, the European Investment Bank (EIB), an autonomous institution funded by EU members, provides financing for investment projects. Its original mandate was to finance projects within the European Union, but its scope was later extended to projects in various developing countries including some in Asia.

VI. Conclusion: Future Prospects of European FDI in Asia

An examination of Europe's presence in Asia has revealed that Europe has lagged behind Japan and the United States as a foreign trade partner and investor in the region. Limited economic ties have prevented Europe from reaping much economic benefit from fast growing Asian economies, and various factors have been seen as barriers to increasing economic relations between the two regions. Furthermore, geographical and cultural distance has hampered active interaction, even as preoccupation with deepening and widening European integration has distracted Europeans from expanding extra-regional relationships. Moreover, the 1997–98 Asian crisis made the region less attractive in the eyes of many firms.

Asia is now recovering quickly from the devastating economic and social impacts of the financial and economic crisis. Although many forecasts predict slower economic growth compared to pre-crisis levels, the Asian economy as a whole is still likely to outperform other regions in the world. In the end, Asia will remain an attractive region for European firms as an export market and as an FDI destination.

There are several signs indicating that Europe's presence in Asia is likely to increase in the future. First, now that economic integration in Europe has been largely consolidated—signified by the introduction of a common currency unit, the euro—European governments and European firms will likely seek to redirect their efforts to increase their ties with non-European regions.

Second, the recent measures implemented to promote Europe's foreign trade and FDI in Asia are expected to have a significant impact in the near future. Indeed, several new cases of large scale FDI from Europe have been undertaken. Among the programs that have been implemented, training and education programs may have a particularly important, if long-term, impact. An improvement in the quality of human resources through training and education programs is a crucial factor for promoting sustained economic growth in Asia, and therefore these programs are universally welcomed by the Asian economies.

Third, successful investment by large European firms is likely to further promote FDI by small- and medium-sized enterprises that carry on business transactions with large firms. European measures to assist SME investment may prove particularly useful, since SMEs lack financial and human resources and thus face difficulty in expanding their overseas activities.

Fourth, we have seen an acceleration of trade and FDI liberalization by Asian economies. They have increased the pace of trade and FDI liberalization for several reasons, most notably due to commitments made at regional and international forums and to the recent economic crisis.[41] Asian economies have also actively pursued unilateral FDI promotion policies beyond their international commitments, ranging from streamlining approval procedures to providing incentives. In fact, they are particularly eager to encourage European FDI as a possible complement to FDI by Japanese and U.S. firms. Asian economies may seek to extract maximum benefits by creating a competitive FDI environment in which investors from different regions compete in the areas of production, sales, technology transfer, and the like.

Finally, the 1997–98 economic crisis has given European firms a chance to increase their presence in Asia. The economic crisis temporarily produced a substantial drop in economic activities in Asia. In particular, the financial and other private sectors suffered seriously because of their huge debt burdens. As a consequence of the weakened financial positions of the financial and manufacturing sectors, Asian firms have not been able to undertake FDI at a pre-crisis level. This state of affairs has allowed European firms to adopt a variety of approaches to increase their presence in Asia, not only through FDI and foreign trade but also through strategic alliances such as technology cooperation. In this regard, effective support from European governments and the EU should prove helpful for European firms.

Notes

1. In this paper "Asia" refers to a large subset of East Asian developing economies, excluding Japan, unless indicated otherwise. This subset includes China, Hong Kong, Indonesia, Korea, Malaysia, Philippines, Singapore, Taiwan, and Thailand.
2. See, for example, World Bank (1993).
3. Both external and internal factors have been seen as the causes of the crisis. One serious external factor was instability in the international financial system, resulting in huge movements of speculative capital on a global scale. As to internal factors, underdeveloped financial and corporate sectors contributed significantly to the emergence of the crisis, as they could not allocate foreign capital efficiently for productive use. Instead, capital was used for speculative purposes such as purchasing land and stocks. As soon as the security of these investments became questionable, speculative short-run capital left Asia vulnerable to currency volatility.

4. "Europe" in this paper refers to the member states of the European Union, unless otherwise noted.

5. Indeed, the European Union has become an economic peer to the United States, not only in terms of GDP but also in population.

6. In the discussion of trade in this section, "Asia" includes the developing economies in Asia and excludes Japan, unless otherwise noted.

7. In 1980–97, Asian exports increased by 5.1 times, while their imports increased by less than 4.7 times.

8. It should be noted that for developing Asian economies the rate of increase for outward FDI stock was particularly high, although the absolute magnitude is still much smaller compared to their inward FDI stock or compared to Japan's outward FDI stock.

9. These shares are computed from the figures in tables 2.1 and 2.2, but they are not shown in the tables.

10. See Kawai and Urata (1998) for a detailed discussion.

11. Kawai and Urata (1996) present evidence of such behavior by Japanese firms.

12. See United Nations (1999) for details.

13. A gravity coefficient indicates the intensity or bias of a bilateral trading relationship by taking into account their importance in world trade. The gravity coefficient is defined as XijX./Xi.X.j, where "Xij" represents exports from region i to region j, and "." indicates the summation across all i or j. Therefore, "Xi." represents total exports of region i, "X.j" total imports of region j, and "X." world trade. The value of unity can be interpreted to mean that the bilateral relationship is neutral, while the relationship is more/less biased when the measure is greater/less than unity.

14. According to Urata (1998a), the gravity coefficients for intra-Europe exports and imports in 1994 were 1.45 and 1.43, indicating that "internal" trade relations are stronger than expected judging from Europe's position in overall world trade.

15. The importance of machinery and transport equipment in European exports to Asia comes into even sharper focus if we note that the share of Europe's exports in other machinery and transport equipment going to Asia is 15.4 percent, significantly higher than the corresponding value for overall European exports at 4.3 percent.

16. While the importance of European trade varies among the Asian economies, the differences are not substantial.

17. An examination of the sectoral composition of European FDI at the global level and in Asia shows that, compared with the rest of the world, European firms have heavily invested in natural resources and capital-intensive sectors, particularly in agriculture, petroleum, and chemicals.

18. European Commission and UNCTAD (1996), Annex table 1.

19. FDI officially originating from Hong Kong includes investment from other sources such as the United States, Europe, and Japan. Given Hong Kong's role

as an intermediary for FDI in other parts of Asia, in particular China, one has to discount the magnitude of intra-Asian, FDI shown in the statistics.

20. European Commission and UNCTAD (1996), Table 1.7.
21. European Commission and UNCTAD (1996), p. 26.
22. European Commission and UNCTAD (1996), p. 30.
23. United Nations (1996), p. 52.
24. Ibid.
25. United Nations (1998, 1999).
26. A lack of recent statistics on FDI in Asia precludes a detailed analysis of the changes in Asian FDI patterns by European firms. However, newspaper reports and anecdotal evidence indicate a significant increase in FDI from Europe in Asia.
27. Bank for International Settlements (1998).
28. Kawai and Urata (1998). Frankel (1997) undertook an in-depth statistical analysis of this bilateral trading relationship and found geographical distance to be an effective deterrent.
29. Boisso and Ferrantino (1997).
30. See Pacific Economic Cooperation Conference (PECC) (1995) and Yamazawa and Urata (1999) for trade liberalization by East Asian economies within the World Bank/IMF, WTO, and APEC frameworks. PECC finds that East Asian economies achieved significant reduction in tariff and non-tariff barriers as a result of the Uruguay round, while Yamazawa and Urata find that in some East Asian economies tariff rates are still quite high for some items despite substantial reduction in average tariff rates.
31. Chia (2000) provides a good discussion of the issues related to AFTA.
32. Kawai and Urata (1998) found a close relationship between FDI and foreign trade in their analysis of bilateral FDI and trade between Japan and its trade and investment partners.
33. Asian FDI regulations in the past and present are nondiscriminatory with one exception. In 1998, the ASEAN Investment Area (AIA) was established for ASEAN members. It provides coordinated ASEAN investment co-operation and facilitation programs and national treatment of all industries, with a temporary exclusion list and sensitive list. The deadline for achieving AIA is set 2010 for ASEAN investors and 2020 for non-ASEAN investors. However, AIA is in the process of being implemented, and therefore cannot be regarded as a factor deterring European FDI in Asia.
34. European Commision and UNCTAD (1996).
35. Urata (1998b) discusses APEC's investment principles.
36. Kawai and Urata (1998).
37. European Commission and UNCTAD (1996), p. 58.
38. This section draws on the information given in the European Commission and UNCTAD (1996).

39. ASEM consists of the fifteen European Union member states and thirteen Asian countries including the ten ASEAN countries, China, Japan, and Korea. See chapter 3 by Ravenhill in this volume for a discussion of ASEM.
40. European Commission and UNCTAD (1996), p.70.
41. Specifically, an agreement was reached among APEC members to complete trade and FDI liberalization by 2010 for developed members and by 2020 for developing members. Liberalization in APEC is nondiscriminatory, and hence barriers to trade and FDI will be removed for firms of all origins. ASEAN members have agreed to liberalize trade among themselves by 2003. Trade and FDI liberalization have also proceeded as a result of the Uruguay Round and creation of the WTO. During that round, rules on intellectual property rights, trade in services, and FDI were formulated and implemented, further reducing the obstacles to global trade and investment.

References

Bank for International Settlement (1998). *The Maturity, Sectoral and Nationality Distribution of International Bank Lending.* Basle, Switzerland: Bank for International Settlement.

Boisso, Dale and Michael Ferrantino (1997). "Economic Distance, Cultural Distance, and Openness in International Trade: Empirical Puzzles." *Journal of Economic Integration* 12 (4), December, pp. 456–484.

Chia, Siow Yue (2000). "Regional Economic Integration in East Asia: Developments, Issues, and Challenges." In *Dreams and Dilemmas,* edited by Koichi Hamada, Mitsuo Matsushita, and Chikara Komura. Singapore: Institute of Southeast Asian Studies.

European Commission and UNCTAD (1996). *Investing in Asia's Dynamism.* Brussels: EC/UNCTAD.

Frankel, Jeffrey A. (1997). *Regional Trading Blocs.* Washington D.C.: Institute for International Economics.

Kawai, Masahiro and Shujiro Urata (1996). "Trade Imbalances and Japanese Foreign Direct Investment: Bilateral and Triangular Issues." In *Asia-Pacific Economic Cooperation: Current Issues and Agenda for the Future,* edited by Ku-Hyun Jung and Jang-Hee Yoo. Seoul: Korea Institute for International Economic Policy.

———(1998). "Are Trade and Direct Investment Substitutes or Complements? An Analysis of the Japanese Manufacturing Industry." In *Economic Development and Cooperation in the Pacific Basin: Trade, Investment, and Environmental Issues,* edited by Hiro Lee and David W. Roland-Holst. Cambridge: Cambridge University Press.

Pacific Economic Cooperation Conference (1995). *Survey of Impediments to Trade and Investment in the APEC Region.* Singapore: Pacific Economic Cooperation Conference.

United Nations (1996). *World Investment Report 1996.* New York and Geneva: United Nations.

———(1998). *World Investment Report 1998.* New York and Geneva: United Nations.

———(1999). *World Investment Report 1999.* New York and Geneva: United Nations.

Urata, Shujiro (1998a). "Regionalization and the Formation of Regional Institutions in East Asia." In *Europe: Beyond Competing Regionalism,* edited by Kiichiro Fukasaku, Fukunari Kimura, and Shujiro Urata. Brighton: Asia & Sussex Academic Press.

———(1998b). "Foreign Direct Investment and APEC." In *Asia-Pacific Crossroads: Regime Creation and the Future of APEC,* edited by Vinod K. Aggarwal and Charles E. Morrison. New York: St. Martin's Press.

World Bank (1993). *The East Asian Miracle.* Oxford: Oxford University Press.

Yamazawa, Ippei and Shujiro Urata (1999). "Trade and Investment Liberalization and Facilitation." Paper presented at the 25th Pacific Trade and Development Conference, June, Osaka, Japan. Forthcoming in *Asia Pacific Economic Cooperation (APEC),* edited by Ippei Yamazawa. London: Routledge.

Chapter 3

Nonmarket Strategies in Asia: The Regional Level

John Ravenhill

I. Introduction

For many Western observers and participants alike, government-business relations in Asia must frequently appear to resemble Winston Churchill's description of Russia: "a riddle wrapped in a mystery inside an enigma." The institutional terrain is disconcertingly unfamiliar. Legal systems are often dissimilar to those at home, or if they bear a superficial similarity, as for instance in some of the former European colonies in Southeast Asia, they operate in practice in a radically different manner. Companies may find their legal rights to be far less secure than they appear on paper. Moreover, considerable "slippage" often occurs between what is agreed at the highest levels of government and what is actually implemented by lower level officials in ministries or in local jurisdictions.

Attitudes toward inward foreign investment have varied substantially across the region. In some Southeast Asian economies, foreign firms have received more favorable treatment at times than their local counterparts; while in some Northeast Asian economies, government attitudes toward foreign investment historically have been unwelcoming at best. Moreover, the financial crises in the region in 1997–98 caused further uncertainties for the foreign business community. Some governments, most notably that of South Korea, have engaged in a dramatic restructuring not only of the relationship between state and business, but also of the major players in the private sector. Elsewhere, especially in Indonesia, political instability has compounded the effects of economic fragility, creating further uncertainty for the foreign business community. The shape of the post-crisis business environment has yet to be finalized in several East Asian countries.

The European business community in East Asia has a wide variety of nonmarket strategies available to it.[1] These range from outright bribery to playing the role of good corporate citizen by sponsoring local charitable organizations, and from seeking strategic alliances with favored local firms or individuals well-placed in politics to lobbying through the local chambers of commerce as noted in Vinod Aggarwal's introductory chapter.[2] The range of strategies is such that it is impossible for one chapter to explore them all. This chapter is intended to parallel that of Cédric Dupont's in focusing on the regional (and transregional) institutions, this time in Asia, through which European companies can pursue their interests through nonmarket strategies.[3] It provides a brief summary of the recent history and evolution of the institutions of the three major regional groupings in which a number of East Asian countries participate: the Association of Southeast Asian Nations (ASEAN), the Asia-Pacific Economic Cooperation (APEC) grouping, and the Asia-Europe Meeting (ASEM).

II. Weak Institutions and Nondiscrimination

The institutional terrain at the regional level in Asia is very different from that in Western Europe. Regional economic cooperation in Asia is for the most part of recent origin (the major exception is ASEAN, which celebrated its thirtieth birthday in 1997) and is only weakly institutionalized. The reasons for the relative newness and weakness of East Asian regional institutions are straightforward. For most of the postwar period, Asia was the principal arena in which Cold War rivalries among the superpowers were fought out in armed conflict. For much of the period (and for some, even today), countries were divided by Cold War conflicts—China and Taiwan, North and South Vietnam, North and South Korea. Elsewhere in the region, the relatively recent date at which some countries acquired their independence from European colonial powers made governments reluctant to enter any regional arrangements that might result in a loss of their sovereignty.

The diversified trade patterns of most East Asian economies have been a further factor inhibiting the formation of a region-wide discriminatory

economic grouping. Whereas Western European economies trade overwhelmingly with one another, East Asian economies continue to depend
heavily on non-Asian markets, primarily the United States, but also the European Union (EU). They have been reluctant, therefore, to support moves
that might do anything to encourage a fragmentation of the global trading
system into rival regional blocs. East Asian governments are generally
among the strongest supporters of the multilateral trading system (even
though China and Taiwan for many years were excluded from the most important global trading institution, the World Trade Organization [WTO]).

East Asian economies are unique among the world's major traders in
not having formed a region-wide discriminatory trading arrangement. Of
the 139 members of the WTO, only three—all East Asian economies:
Hong Kong, Japan, and South Korea—are not currently parties to any discriminatory regional trading arrangement.[4] Because of the concern to
maintain access to external markets, especially that of the United States,
East Asian economies historically have preferred a transregional trading
agreement (APEC) to one confined to the Western Pacific Rim. This preference was demonstrated most clearly in 1989 when, faced with the im-
minent establishment of APEC, Malaysia's Prime Minister Mahathir
Mohamad proposed that an East Asian Economic Grouping be created.
The proposal met with hostility not only from the United States but also
from the leaders of several other East Asian economies because they feared
an East Asian discriminatory trading arrangement would threaten their access to North American markets. In a face-saving move, East Asian leaders
agreed that the proposed grouping would be realized as an East Asian Economic Caucus that would meet within APEC rather than as a separate entity established in opposition to this broader grouping.

In recent years, however, East Asian economies have engaged in more
intensive discussions about regional economic integration. The East Asian
Economic Caucus was cemented by the establishment of a second transregional grouping, ASEM, where the Asian component is essentially those
East Asian countries that Dr. Mahathir envisaged as members of his proposed East Asian Economic Grouping.[5] And, in the wake of the financial
crises and Asian governments' resentment at what they perceived as an un-
helpful response from Western governments, East Asian countries have
consulted more frequently with one another, primarily through meetings
associated with ASEAN fora. The most important of these has been the
"ASEAN Plus Three" (ASEAN plus China, Japan, and Korea) grouping,
which began in 1998 to hold annual meetings of heads of state following
the ASEAN summit. More regular consultations among the East Asian

economies notwithstanding, the possibility of a discriminatory trading bloc emerging remains remote given the diversity of interests of the East Asian economies.[6]

The two organizations that have the most comprehensive membership of East Asian economies, APEC and ASEM, are transregional rather than regional in membership. The former includes the members of the North American Free Trade Agreement (NAFTA), the Australia-New Zealand Closer Economic Relations (CER) agreement, Peru and Chile as well as East Asian economies, while the latter links a smaller group of East Asian states with the European Union. The only exclusively East Asian regional economic agreement, ASEAN, is confined to the Southeast Asian region—currently with ten members.[7]

The relative institutional weakness of these groupings is implicit in their titles. APEC has failed to move beyond its depiction by the former Australian foreign minister, Gareth Evans, as "four adjectives in search of a noun." A move to substitute "community" for "cooperation" in the grouping's title was rejected by its member economies because the term implied more institutionalization than they wished to see. ASEM is but a "meeting." And even the most institutionalized of the groupings, ASEAN, is only an "association" of states. ASEAN, as the oldest and most established of the arrangements, is an appropriate starting point for understanding the reasons why regionalism in Asia is so underdeveloped relative to Europe, and reveals the implications of this lack of institutionalization for European business.

ASEAN

ASEAN had its origins in British decolonization in Southeast Asia in the late 1950s, and in the formation and quick dissolution of the Federation of Malaysia in 1963 and 1965 respectively. Relations between Malaysia and Singapore were soured by the dissolution of the Federation; the Malaysian government meanwhile was in armed conflict with the Indonesian government of President Sukarno. More broadly, the Southeast Asian states faced an uncertain strategic climate as war escalated in Vietnam and doubts grew about the continuing commitment of the U.S. government to this struggle. Essentially, ASEAN was created as a confidence-building mechanism among the five founding member states in 1967. The grouping was intended to not merely help diminish the likelihood of conflict among the member states, but also strengthen their bargaining position (both in economic and security matters) vis-à-vis other countries. Although economics provided the ostensible grounds for cooperation, security issues were the primary concern of ASEAN leaders at this time—and, arguably, have remained so ever since.

In its first twenty-five years, ASEAN achieved very little in the economic sphere, in marked contrast with the European Community. Several factors contributed to this poor record. Economic relations with other member economies were of relatively minor consequence to ASEAN countries: Southeast Asian economies were little integrated with one another. Trade with ASEAN partners as a whole seldom exceeded 20 percent of the member economies' total trade; most of this relatively small percentage comprised trade between Singapore and the other economies. Member governments pursued divergent economic policies that ranged from the low tariff and open foreign investment regime of Singapore to the highly protectionist policies of Indonesia and Thailand.

Moreover, the programs for economic cooperation contemplated by ASEAN in the 1970s were heavily influenced by the developmental thinking of the time that favored industrial planning, an approach unpopular with the more industrially advanced economies of the region. The Basic Agreement on ASEAN Industrial Projects signed in Kuala Lumpur on March 6, 1980, for instance, provided for the allocation of five major industrial projects among the member economies. Little consultation occurred with the private sector in identifying the proposed projects and determining to which country they should be allocated. Only two of the original five projects were implemented (both of which had previously been put forward as national projects); the other proposals foundered because of squabbles among the member states, or because of domestic disagreements over the location of the project.

ASEAN's subsequent ventures into regional industrial cooperation were only marginally more successful. They did, however, have greater input from the private sector. In 1976 the ASEAN Automotive Federation put forward a scheme for ASEAN Industrial Complementation (AIC), building on a proposal originally made by the Ford motor company in 1971. The objective was to facilitate exchange of components and realize economies of scale in a region–wide market. The AIC program was eventually implemented in 1981. The project had only limited success, however, owing to disagreements among the member states and to the incompatibility of parts sourced from different producers. A revamped version, titled Brand-to-Brand Complementation (BBC), was implemented in 1988, largely at the behest of multinational automobile companies (but Indonesia refused to participate, preferring to provide protection for its national car scheme).

The last of the measures to promote joint industrial production was the ASEAN Industrial Joint Venture agreement (AIJV). Originally proposed by the ASEAN Chambers of Commerce and Industry in 1980, the agreement was signed in 1983. To enjoy AIJV status, which provided for a 90 percent preference in tariffs, projects had to have the participation of two or more ASEAN countries, and satisfy rules of origin requirements. The private sector showed little interest in the arrangement, however, in part because ASEAN did little to promote the arrangements, but also because of the long delays before state bureaucracies approved proposals.

In the trade sphere, ASEAN was similarly unsuccessful. Its Preferential Trading Arrangements (PTA), introduced in 1977, were notorious for the exemptions that governments claimed from obligations to lower tariffs on goods produced competitively by fellow members, and for granting preferences on intraregional trade for goods not actually produced within the region.[8] By the end of the 1980s, trade in the nearly 16,000 products that nominally enjoyed preferences under the PTA amounted to less than one percent of intra-ASEAN trade.[9]

Underlying the lack of progress on economic cooperation was a fundamental institutional weakness. ASEAN governments have been determined that they should not lose control of the cooperation process to any regional institution. There would be no question of the majority overriding a recalcitrant minority in pushing for deeper cooperation. The "ASEAN Way" dictated that progress was to occur based on consensus among the member states; in effect, cooperation has been held hostage to a lowest common denominator effect. Accordingly, unlike the European Community, member states did not endow ASEAN with a powerful secretariat. Rather, jealously guarding their sovereignty, they chose to keep the ASEAN Secretariat weak and underfunded. The primary responsibility for servicing the organization rested with the bureaucracies of the individual member states. ASEAN consequently lacked an institution equivalent to that of the Commission of the European Communities that enjoyed sufficient autonomy to take initiatives to drive the cooperative process forward.

ASEAN member states arguably only began to take regional economic cooperation seriously in the late 1980s. The most important precipitant here was the emergence of China as a significant competitor for foreign direct investment (FDI). ASEAN governments increasingly realized that a fragmented Southeast Asian market further diminished their attractiveness to foreign investors compared with the Chinese behemoth. Some evidence exists that foreign, particularly Japanese, corporations exerted increasing pressure on Southeast Asian governments to unify their markets. Another factor influencing ASEAN governments to deepen their cooperation at this time were the proposals to establish a transpacific regional economic grouping, proposals that came to fruition in November 1989 with the first meeting of APEC. ASEAN member states feared that their organization would be sidelined if it failed to pursue an agenda that addressed the need for closer regional economic cooperation. Doubts as to whether the Uruguay Round of the General Agreement on Tariffs and Trade (GATT) negotiations would reach a successful conclusion also spurred ASEAN discussions. The prospects for closer regional economic cooperation were also enhanced by changes in the economic structures of Southeast Asian economies that took place in the 1980s. Of particular significance was the rapidly growing share of manufactured goods in their exports, driven in large part by substantial inflows of FDI in this period. Less dependent on primary products than in the past, the ASEAN economies increasingly had the potential to engage in increased trade with one another.

In response to the uncertain external environment, ASEAN heads of state and government at their fourth summit meeting in January 1992 agreed to establish an ASEAN Free Trade Area by adopting a Common Effective Preferential Tariff (CEPT). All tariffs on intra-ASEAN trade in manufactures and processed agricultural products were to be lowered to the range of 0 to 5 percent within fifteen years beginning in January 1993, a period subsequently reduced to nine years.[10] The member states also agreed to remove all quantitative restrictions on imports once products begin to enjoy concessions under the CEPT, and to remove all other non-tariff barriers within five years of first receiving these concessions. The Agreement also provides for the creation of a ministerial-level ASEAN Free Trade Area (AFTA) Council to supervise the implementation of the CEPT. At its December 1992 meeting, the Council agreed to rules of origin under the CEPT. Products will be eligible for the CEPT if 40 percent of the value has been added within one ASEAN country, or if the cumulative ASEAN content reaches at least 40 percent.[11]

Rather than following the time-consuming procedure under the PTA of identifying individual products at the eight or nine digit tariff levels for liberalization, the CEPT covers all manufactured products except those specifically excluded. The publication of a timetable for tariff liberalization further increases the pressure on governments to proceed with tariff reductions. Out of over 15,000 tariff lines identified by the member states, more than 90 percent are included in the CEPT, accounting for more than 81 percent of the current value of intra-ASEAN trade.

To accommodate the creation of the free trade area, the member states agreed at the fifth ASEAN summit in Bangkok in December 1995 to phase out the existing industrial cooperation schemes (BBC and AJIV) and replace them with a new arrangement. The new ASEAN Industrial Co-operation Scheme provides tariff rates of zero to five percent with immediate effect to firms that conduct approved manufacturing activities across ASEAN boundaries. In addition, in 1998, the grouping established an ASEAN Investment Area, an arrangement intended to facilitate foreign direct investment and capital movement among the member states.

Although AFTA embodies a far more substantial commitment from the member states than did the PTA, it has several significant weaknesses. These fall into two principal categories: lack of specificity of provisions and weak institutional structure and capabilities. The lack of specificity of the agreement reinforces the view that AFTA was as much a political gesture as an economic instrument whose ramifications had been thoroughly explored prior to signature. Contrast, for instance, the dozen or so pages of the Singapore Declaration with the more than one thousand pages of the North American Free Trade Agreement. The consequence of the lack of detail in the Singapore Declaration is that many key issues are unresolved and are having to be negotiated at the same time as the member states are supposed to be implementing tariff reductions. The lack of specificity in the treaty has led some to dub AFTA "Agree First, Talk After." Moreover, the expansion of ASEAN to ten member states has complicated implementation of AFTA with new members being given additional time to fulfil their obligations.

Two of the key areas on which much work remains to be done are non-tariff barriers and rules of origin. Nontariff barriers are at least as serious an impediment to the growth of intraregional trade in ASEAN as are tariffs. Many ASEAN states deploy a wide range of nontariff barriers, including domestic standards requirements, discriminatory government procurement practices, licensing for the distribution of certain products, local content and counter-purchase requirements, and the often arbitrary application of customs classification and valuations.[12] No comprehensive survey of the extent of nontariff barriers and their effects on intra-ASEAN trade had been undertaken by the time of the Singapore Declaration. Member states agreed to provide information to the ASEAN Secretariat by the end of 1993 about the nontariff barriers they impose, but this information has yet to be made public. The difficult job of negotiating and monitoring the removal of these barriers remains to be done; the Singapore Declaration provides no formal framework for accomplishing this.

A similar lack of specificity is found in the rules of origin and safeguard provisions. Although the rules adopted in the agreement on the CEPT signed at the time of the Singapore Declaration are straightforward, they may well prove to be unworkable. In particular, the valuation of local content can be easily manipulated by transfer pricing, the under-invoicing of components purchased from extraregional suppliers. As Singapore and Brunei are free ports, other ASEAN states are concerned that goods originating outside the region will be diverted through these ports and relabeled as ASEAN products. The ASEAN Federation of Textile Industries and the ASEAN Chambers of Commerce and Industry have suggested that a substantial transformation requirement, similar to that used in NAFTA, will be necessary if the rules of origin are not to be abused. A requirement of this type, however, would require a complete rewriting of the present rules. The CEPT's safeguard clause, meanwhile, fails to specify the conditions that would justify its deployment, or for what period safeguard measures can be utilized, leaving it open to possible abuse.

In any free trade area, the complexities are such that effective institutions and measures for dispute adjudication and rule interpretation are required. This is especially the case in AFTA given the lack of specificity of the agreement. Yet here again the provisions of the agreement are particularly weak. ASEAN has a long tradition of preferring political to administrative or juridical arrangements for the settlement of disputes. The "ASEAN way" of decisionmaking by consensus has served as a rationalization for the unwillingness of member states to transfer any decision-making authority to regional institutions.[13] Key decisions are made by meetings of the economic ministers of member states, to whom disputes are referred if they cannot be resolved by meetings of senior national economic officials. If economic ministers cannot resolve disputes, then they are referred to heads of state. ASEAN was embarrassed when two of its member states, Malaysia and Singapore, appeared in the very first case to be heard by a WTO dispute settlement panel.[14] In response, ASEAN established its own dispute settlement procedures in November 1996. By the end of 1999, however, no state had referred a dispute to them.

As part of the negotiations for AFTA, the member states agreed to strengthen the ASEAN Secretariat. Recruitment to the Secretariat would henceforth be on the basis of open competition from nationals of member states rather than secondment from national ministries, although such recruitment would continue to be subject to national quotas. The number of professionals within the Secretariat was more than doubled to thirty-five. The Secretariat remains responsible not only for the implementation of AFTA, but also for the grouping's external relations. The contrast between the resources available to the ASEAN Secretariat and those enjoyed by the Commission of the European Union is marked.

Even after more than thirty years of interstate cooperation, ASEAN remains a weak international institution. Decisionmaking power remains concentrated in the hands of officials of member states. Doubt exists whether ASEAN has an international legal personality and treaty-making capacity: unlike the European Union, where one signature, the President of the Council of Ministers, represents the Union, treaties involving ASEAN usually have signatures from each of the member countries. The Secretariat continues to be understaffed and lacking in any capacity to take initiatives. And ASEAN's measures to promote economic cooperation remain largely ineffective—even in comparison with those of other groupings of less developed economies. Its progress in implementing its free trade area, for example, pales in comparison with the rapid implementation of the MERCOSUR agreement among Brazil, Argentina, Uruguay, and Paraguay. Trade between ASEAN member economies in the 1990s grew only slightly more than one percent more rapidly than their trade with other countries. And even this modest figure probably overstates the importance to intraregional trade of ASEAN trade preferences: a study by the ASEAN Secretariat estimated that only 1.5 percent of intra-ASEAN trade used the certification of origin required to attain preferential treatment.[15] Few companies have made use of ASEAN's industrial cooperation procedures because of bureaucratic delays in decisionmaking.

What, then, are the implications for European firms' pursuit of nonmarket strategies in Southeast Asia? A first point is that companies cannot have any confidence that agreements on trade reached at the regional level in ASEAN will actually be implemented by the member economies. Even if members move to implement AFTA on schedule (and this is by no means assured, several countries having raised tariffs in the wake of the 1997–98 financial crises), the actual tariff prevailing on a product in the various member states may be anywhere from 0 to 5 percent. Variations in tariffs and in the timetable for their reduction across countries are likely to cause difficulties if not confusion for business. As noted above, the grouping has yet to work out many other details of the implementation of the free trade arrangement; it is quite likely that ambiguity on many of these issues will persist. And, as the deadline for implementation of the agreement draws closer, some countries have sought exemptions from liberalization for products regarded as politically sensitive: the most notable example has been Malaysia's determination to delay reduction in the tariffs protecting its domestic auto producers. Because of political sensitivities, dispute resolution is likely to remain a matter of politics rather than juridical or administrative procedures.

A second implication concerns where European firms might concentrate their nonmarket strategies. Effective power within ASEAN continues to rest with the member states. Meetings of the senior officials of the member states are the day-to-day decisionmaking body. The Secretariat has at best a very limited role in proposing initiatives; it has lacked the capacity to develop detailed proposals or to monitor their implementation. Therefore, scarce lobbying resources are more likely to generate a positive return if invested at the national government level, especially in those countries (most notably Malaysia and Singapore) that possess the more efficient public services in the region. Other national governments of course may also be mobilized when firms perceive an important economic issue to be at stake. Also, though multinational companies may have an interest in rationalizing production across the ASEAN region as a whole, the current ASEAN Industrial Cooperation scheme is subject to too many bureaucratic delays to be attractive to corporations in a fast-moving industry. Again, the principal remedy seems to lie through lobbying national governments because it is they rather than the ASEAN Secretariat that have been the principal cause of the delays.

If companies wish to work at the regional level, then the ASEAN Chambers of Commerce and Industry (ASEAN-CCI) may be an effective lobby group. The ASEAN-CCI was set up more than a quarter of a century ago to explore greater private sector participation in the ASEAN policy formulation process. The membership of ASEAN-CCI is made up of one national private sector body representing commerce and industry from each ASEAN member country. Corporate associate membership is available to any company located in ASEAN region that would like to participate in the ASEAN-CCI activities and agrees to comply with the ASEAN-CCI constitution.

APEC

The Asia-Pacific Economic Cooperation grouping, established in 1989, links East Asian economies with others around the Pacific Rim. Although Russia, ostensibly far more a European than an Asian economy, became a member of APEC in 1998, the grouping has refused to widen membership beyond territories with a Pacific Ocean littoral. APEC has persistently declined to provide any official status for the European Union in its meetings. Unlike other regional trade arrangements, however, APEC has committed itself to the principle of "open regionalism." Although this term has some ambiguity,[16] the essence is that APEC members will apply to all their trading partners the advantages that they give to other APEC members. In other words, APEC is to operate on a nondiscriminatory basis rather than seeking exemption from the most-favored-nation requirements of the WTO under the provisions of Article XXIV of GATT.[17]

APEC committed itself at the Bogor summit in 1994 to achieving free trade among member economies by the year 2020 (for industrialized economies the deadline is 2010). Since that agreement, however, considerable disagreement has arisen among member economies over the implications of the commitment. Whereas the governments of North American and Oceanic member states have viewed the commitment as legally binding, many of those from East Asia see it as an indicative target, with compliance being voluntary. APEC economies have produced individual "action plans" to document their progress toward removing tariff and nontariff barriers, but these have seldom gone beyond commitments that they have made in other fora, especially the Uruguay Round of GATT talks. At the APEC leaders' meeting in Kuala Lumpur in 1998, attempts to promote more rapid trade liberalization through targeting nine industrial sectors for immediate elimination of tariffs foundered when the Japanese government refused to agree to negotiate within APEC on two of the sectors, fisheries and forests. This program of "Early Voluntary Sectoral Liberalization" was referred to the WTO (which has yet to act on it).

Although trade liberalization has been at the forefront of public interest in APEC since the Bogor declaration, it is but one of three "pillars" on which APEC rests. The others are trade facilitation and economic and technical cooperation. The latter has been of particular interest to the less-developed members of the grouping and has been championed by the Japanese government (which has never given high priority to the trade liberalization agenda in APEC). APEC has three committees, an ad hoc policy group, and ten working groups devoted to promoting economic and technical cooperation in various sectors.[18] These various committees are engaged in over 350 projects that together involve most government ministries in the member economies. Trade facilitation, however, is perhaps the area where APEC may yet make its most significant contribution to the interests of international business. APEC is working in association with business groups on various measures to reduce the transaction costs of trade among its member economies. If it is successful in harmonizing customs schedules or testing requirements, APEC will have taken a significant step to facilitating commerce across all of its member economies.

In its institutions and decisionmaking structures, APEC resembles ASEAN. This is no coincidence. Securing the participation of ASEAN members had been a longstanding obstacle to the realization of a transpacific economic grouping. Only by assuring ASEAN governments that decisionmaking would be by consensus and that there would be no question of a majority overriding a minority within the grouping was ASEAN participation attained. Like ASEAN, therefore, APEC can only proceed at the pace of the least willing participant. Also like ASEAN, APEC has eschewed the creation of a large secretariat that might act independently to take initiatives on regional cooperation. The small secretariat based in Singapore does little more than coordinate the activities of APEC's various groups.

Rather than attempting to legislate, APEC relies primarily on socialization and peer pressure: the hope that exposing all participants to the force of pro-liberalization ideas, plus the desire of not wanting to be perceived as a spoiler, will induce cooperative behavior. Where APEC has reached agreement, as in its investment principles, the rules are nonbinding. As with ASEAN, the consequence is that business cannot be sure that member economies will actually implement any agreement reached at the regional level. APEC has no mechanisms for dispute resolution or arbitration. The nonbinding character of its agreements and the opportunity for governments to provide their own interpretation of the obligations that they have undertaken preclude the development of such mechanisms.

APEC has repeatedly proclaimed its pro-business orientation. Joan Spero, a former U.S. Undersecretary of State, commented that "APEC is all about business." Business is the only interest group that has an institutionalized relationship with APEC—through the APEC Business Advisory Council (ABAC). Members of the Council (three from each country) are nominated by the governments of the member economies, casting some doubt as to their total independence from government influence. ABAC has, however, been outspoken on occasion at what it perceives to be insufficient progress within APEC toward the goals of more liberalized trade and investment flows.

For the most part, members of ABAC come from locally-owned companies. For several of the smaller economies, however, representatives have been appointed from the local subsidiaries of multinationals, including European companies.[19] So despite APEC's not including any European Union member state and refusing a formal link with the EU, European business enjoys a channel for influencing APEC's deliberations through the Business Advisory Council, though the effectiveness of this channel is debatable. ABAC is a consultative body and is not directly engaged in decisionmaking.. Its influence lies in moral suasion rather than in any control its members have over outcomes.

ABAC is much like APEC itself: very much a glass-half-full, glass-half-empty matter in which the interpretation of its usefulness depends on individual perspectives on the utility of persuasion and deliberative process. Critics of APEC focus on the lack of legally binding resolutions, the slow pace of liberalization, and the distant deadline for calling governments to account for the trade liberalization commitments they made at Bogor in 1994. APEC's proponents, in contrast, argue that only an organization that has little institutionalization and operates on the basis of consensus would be acceptable to most East Asian governments. The increased willingness of East Asian governments to open their economies in the 1990s, they assert, is testimony to the power of ideas and of peer pressure that operated first within the Pacific Economic Cooperation Conference. APEC now builds on and extends this collaborative process.

ASEM

The Asia-Europe Meeting is the most recent of the transregional groupings that East Asian states have entered and thus, not surprisingly, is even less institutionalized than ASEAN or APEC. Unlike APEC, which built on a history of more than a quarter of a century of non-official cooperation across the Pacific, ASEM had few institutional links between Europe and Asia on which to capitalize—beyond the dialogue between the EU and ASEAN.

ASEM arose in part from European concerns that European companies were missing out in the rapidly growing East Asian markets, and that the EU was not exerting influence in the region proportional to its strength in the global economy. These concerns were manifested in calls by the European Commission in the early 1990s for a new Asian strategy. The driving force on the East Asian side was the desire to seek another alliance partner that would help offset what they perceived as the excessive leverage the United States enjoyed within APEC (at a time when Washington was rigorously pushing a trade liberalization agenda within the organization).[20] Lacking a region-wide economic grouping of their own, East Asian governments saw a new transregional relationship with Europe as a means to increase their bargaining leverage within their existing transregional relationship with North America.

Like APEC, ASEM is organized around three "pillars," in this instance the fostering of political dialogue, the reinforcing of economic cooperation, and the promotion of cooperation in social and cultural fields. ASEM eschewed the establishment of new formal institutions—hence the emphasis on "meeting." It has no secretariat; the bureaucracies of its member states service its meetings. ASEM holds biannual summits, the third of which took place in Seoul in October 2000, and more frequent meetings at ministerial and senior official levels.

Also like APEC, ASEM puts a great deal of emphasis on facilitating business relations. ASEM's foreign ministers, in a meeting in Singapore in February 1997, noted the need "for a greater private sector role in new programs of economic cooperation." To promote improved relations between business in the two regions, ASEM established an Asia-Europe Business Forum. This body in turn organizes an Asia-Europe Business Conference, which meets annually with working groups devoted to issues such as infrastructure and financial services. ASEM members have adopted an Investment Promotion Action Plan and a Trade Facilitation Action Plan, both designed to break down barriers to commerce between the two regions.

Despite the emphasis that ASEM has placed on business, observers have suggested that business has shown little interest in the new grouping. ASEM, Ku-Hyun Jung and Jean-Pierre Lehmann assert, is largely "symbolic."[21] They note that "there is not much that ASEM can actually 'do' in relation to business and the economies whether of the two regions separately or in the interaction between the two." ASEM, they conclude, is best regarded as a channel for information flow and a possible watchdog over state behaviors.

III. Conclusion

The landscape of regional economic cooperation in East Asia has changed dramatically since the mid-1980s. At that time, only one (sub)regional organization, ASEAN, was in existence—and its achievements in the economic sphere were minimal. In the following decade, new cooperative arrangements blossomed in security and economic affairs. But no new economic organization materialized that was confined to East Asia. Instead, both of the two major new groupings, APEC and ASEM, were transregional, linking East Asia with other parts of the world. At the turn of the century, the increasing range of activities associated with the ASEAN Plus Three grouping suggests that a significant switch of government interest and energies is under way, with the possibility of institutionalized cooperation on an East Asian basis emerging. Moreover, disappointment at the inability of APEC to make significant progress on its trade liberalization agenda and at the failure of the WTO ministerial meeting in Seattle, coupled with increasing concern at the success of regional collaboration in the Americas and Europe, has prompted some Asian governments to show a new interest in bilateral trading agreements.[22] Collaboration among East Asian governments is evolving rapidly and is currently attracting greater government interest than do the transregional groupings. At the time of writing, however, collaboration through bilateral agreements or an ASEAN Plus Three basis is but a set of proposals under consideration by study groups.

The three existing major economic cooperation arrangements that include East Asian governments operate in a very different manner from the European Union. The institutional landscape is much simpler. No supranational institutions exist. Regional secretariats are weak, generally lacking the authority to take independent initiatives. To some degree, the weakness of regional institutions involving East Asian economies might be seen as advantageous to business. The existing regional secretariats are unlikely to propose measures that aim to constrain the activities of business. Indeed, business holds a privileged position: it is the only organized interest group to enjoy an institutionalized relationship with the three regional bodies. The absence of any institution of popular representation equivalent to the European Parliament robs potential opponents of business of a forum through which to promote their views.

Yet if the lack of institutionalization of the regional bodies in some ways may increase the certainty of the environment in which business operates, in other ways it definitely undermines it. The absence of legally binding agreements, the acceptance of individual member states' interpretations of the obligations they have assumed, and consensus-based decisionmaking all contribute to uncertainty not only over the outcomes of regional deliberations, but also over whether any regional agreement will ever be implemented by member states. Whether the new bilateral arrangements currently being negotiated will provide any greater certainty remains to be seen.

It would be incorrect to dismiss the three major regional bodies discussed in this chapter as mere talking shops. The forums are useful not only for the exchange of information (and thus the enhancement of transparency), but also for monitoring the behavior of governments and for putting pressure on recalcitrants. Business may well decide, however, that it can pursue nonmarket strategies more effectively at the level of national governments in Asia or in Europe rather than within these weak regional institutions.

Notes

1. By East Asia, I mean both Northeast Asia and Southeast Asia.
2. See chapter 1 by Aggarwal in this volume.
3. See chapter 4 by Dupont in this volume.
4. In February 1999 Japan and South Korea announced their intention to explore the possibility of linking their economies through a preferential trading arrangement. Few commentators, however, believe that a (comprehensive) bilateral preferential trade relationship between these two countries will be established in the foreseeable future. The first bilateral trade agreement that Japan will sign is likely to be with Singapore; the two sides negotiated the framework for this treaty in 2000.
5. ASEM, unlike APEC, excludes Hong Kong and Taiwan because of the sensitivities of the Chinese government to the inclusion of these territories in a forum whose discussions range beyond economic issues.
6. Ravenhill (2000).
7. The five original members of ASEAN were Indonesia, Malaysia, the Philippines, Singapore, and Thailand. Brunei joined in 1984. Vietnam was added in 1995, Burma and Laos in 1997, and Cambodia in 1999.
8. See Ravenhill (1995), pp. 850–866.
9. Yam, Heng, and Low (1992), pp. 309–332.
10. ·ASEAN countries are free to determine the tariffs they impose on imports from outside the group. That is, AFTA is not a customs union. Although member economies are supposed to complete AFTA by 2002 (for the six original signatories of the treaty), the grouping has set a deadline of 2015 to achieve zero customs duties for the six original signatories and 2018 for the four more recent members. The ASEAN Secretariat estimates that by the end of 2002 approximately 70 percent of trade between the six original signatories will be subject to zero tariffs.
11. For the purposes of cumulation, the rules of origin are liberal in that they treat a component that reaches the 40 percent local origin rule as being 100 percent locally produced. AFTA is a discriminatory arrangement. The AFTA Secretariat claims, however, that the inclusion of most trade in manufactures makes AFTA compatible with the WTO's regulations on regional free trade areas.
12. See Ibrahim and Isa (1987), pp. 74–96; Tin (1987), pp. 97–113; and Pacific Economic Cooperation Council (1995).
13. See Thambipillai and Saravanamuttu (1985), pp. 3–28.
14. A complaint by Singapore against Malaysian restrictions on the import of certain chemicals was resolved in true ASEAN style when Singapore withdrew the complaint in July 1995.
15. Preferential tariffs are lower than MFN rates in less than one third of the tariff lines. See Teh Jr. (1999).

16. "Open regionalism" is a matter of controversy because some members, most notably the United States, have insisted that they will not offer concessions to nonmembers unless reciprocity is provided.

17. Article XXIV legitimizes departure from the most favored nation requirement providing regional trade arrangements meet specified criteria. For a discussion of APEC's evolution see Aggarwal and Morrison (1998).

18. See the APEC website: http://www.apecsec.org.sg.

19. These include directors of the Brunei subsidiary of Shell, and of the Union Bank of Switzerland, East Asia.

20. See Camroux and Lechervy (1996), pp. 441–452; Bridges (1996), pp. 204–218.

21. See the APEC website.

22. Besides the negotiations discussed above between Japan and Singapore, and between Japan and South Korea, the government of Singapore is also negotiating bilateral arrangements with the United States, New Zealand, Australia, and Chile. New Zealand and Australia are also reported to be interested in negotiating bilateral deals with the United States.

References

Aggarwal, Vinod K. and Charles Morrison (1998). *Asia Pacific Crossroads: Regime Creation and the Future of APEC.* New York: St. Martin's Press.

APEC web site: http://www.apecsec.org.sg.

Bridges, Brian (1996). "Western Europe and Southeast Asia." In *Southeast Asia in the New World Order: The Political Economy of a Dynamic Region,* edited by David Wurfel and Bruce Burton. New York: St. Martin's Press.

Camroux, David and Christian Lechervy (1996). "'Encounter of a Third Kind?': The Inaugural Asia-Europe Meeting of March 1996." *Pacific Review* 9 (3), pp. 441–452.

Ibrahim, Rahman and Mansor Md. Isa (1987). "Non-Tariff Barriers to Expanding Intra-ASEAN Trade: Malaysia's Perceptions." *ASEAN Economic Bulletin* 4 (1), pp. 74–96.

Pacific Economic Cooperation Council (1995). *Survey of Impediments to Trade and Investment in the APEC Region: A Report by the Pacific Economic Cooperation Council for APEC.* Singapore: Pacific Economic Cooperation Council.

Ravenhill, John (1995). "Economic Cooperation in Southeast Asia: Changing Incentives." *Asian Survey XXXV,* pp. 850–866.

———. (2000). "APEC Adrift." *Pacific Review* 13 (1).

Teh, Robert Jr. (1999). "Completing the CEPT Scheme for AFTA." Paper presented at conference, Beyond AFTA and Towards an ASEAN Common Market, Manila (October).

Thambipillai, Pushpa (1985). "ASEAN Negotiating Styles: Asset or Hindrance?" In *ASEAN Negotiations: Two Insights,* co-authored by J. Saravanamuttu. Singapore: Institute of Southeast Asian Studies.

Tin, Ooi Guat (1987). "Non-Tariff Barriers to Expanding Intra-ASEAN Trade." *ASEAN Economic Bulletin* 4 (1), pp. 97–113.

Yam, Tan Kong, T. M. Heng and L. Low (1992). "ASEAN and Pacific Economic Cooperation." *ASEAN Economic Cooperation* 8 (3), pp. 309–332.

Chapter 4

Euro-Pressure:
Avenues and Strategies for
Lobbying the European Union

Cédric Dupont

I. Introduction

The creation of the European Community (EC) in 1957 marked a significant change in the nonmarket environment of firms in Europe. Regulatory authority has since evolved toward a fragmented, multilevel structure. Market size, regulation, and development have significantly depended upon the behavior of several political actors at the national, supranational, and subnational levels. Whereas the multiplicity of political actors has long existed, its significance for firms has largely increased since the mid-1980s. Deeper integration, notably through the implementation of the single market program and the advent of a single European currency, has changed the scope and depth of the legal framework of the European Union (EU), increasing the authority of supranational institutions. As a result, power sharing among member states, the European Commission, and the European Parliament has become more diffuse, leading firms to diversify their targets for influence and pressure.

In this new context, lobbying has mushroomed in Brussels as firms have frantically tried to get involved in EU decisionmaking. Firms have developed lobbying strategies in relation to the various stages of the policy cycle (policymaking, implementation, and adjudication).[1] However, increased lobbying has not yet produced corresponding political influence. Facing a complex and evolutionary polity, firms have pursued as many EU roads as possible and have done so without great care for coherence and knowledge, and thus often with little efficiency. Analysts, for their part, have gradually paid higher attention to this domain of activity. They have tried both to document the practice of firms in various sectors of activities and to generate theoretical insights on patterns of lobbying at the EU level.[2] Accordingly, there now exists a meaningful body of scholarly work on firms' nonmarket activities in Europe. However, few solid theoretical findings have come out of this literature. There has been little scientific accumulation from the large number of specific case studies, and there is a lack of synthesis in trying to account for the strategic nonmarket choices of firms. Even the most developed insights fall short of providing systematic knowledge. For example, some authors have developed a useful matrix conceptualizing business access to European channels.[3] This matrix considers four elements: the target, the organizational form of corporate activity, the route, and the voice. The potential of this framework, however, is limited by the fact that the authors provide little, if any, precise guidelines on how firms navigate inside the matrix. On a different note, work on patterns of governance inside the EU, analyzing the evolving forms of associations in both business and labor, provides few specific findings for the strategies of individual firms.[4]

This chapter builds on existing work and seeks to develop a synthetic account of firms' preferences regarding their choices at the stages labeled positional and strategic in the introduction to this volume. I develop a decision map for firms engaged in lobbying activities at the EU level, which assigns a choice of and route to the target as well as the organizational form for sixteen different situations. In constructing the decision map, the chapter begins with a general discussion of the nonmarket context of European firms. The first section addresses the complexity of the EU polity and its evolutionary nature. The second section develops the decision map; it first discusses the choice of targets, routes and organizational forms separately, and then combines the three dimensions into a synthetic analytical framework. The third section provides an application of the framework to examine the efforts of European firms to penetrate Asian markets.

II. The Institutional Context of European Firms: Evolutionary Complexity

The 1957 Treaty of Rome, which created the European Economic Community (EEC), laid down a large agenda of trade liberalization, trade facilitation, and economic cooperation. From the ultimate perspective of preserving and strengthening peace and liberty, member countries stated their willingness to promote economic development and growth and to reduce disparities between them. In line with these broad objectives, they agreed on a timetable for the implementation of a common market for industrial and agricultural products with a common external tariff and harmonization of fiscal and economic policies. This included harmonization of policies in agriculture, energy, transport, and competition, and the development of fiscal transfers to promote development in poorer regions.

The Treaty of Rome created an elaborate institutional structure for the achievement of the broad agenda of the organization. The cornerstones of this structure are three supranational organs: the European Commission (the Commissioners and the Administration), the European Court of Justice (ECJ), and the European Parliament—and one intergovernmental forum—the Council (under its various forms). With respect to relative prerogatives, the text of the Treaty of Rome tended to tilt the balance of power in favor of supranationalism. In addition to the creation of three main supranational organs, selective majority decisionmaking inside the Council of Ministers was intended to boost institutional momentum and power. In the early years of the Community, however, unanimous decisionmaking was the rule—in particular after the so-called Luxembourg compromise of 1966 gave veto power to every individual member. The compromise transformed the Community into a de facto intergovernmental body with limited supranational power, belonging almost exclusively to the Commission. The latter gradually benefited from the help of the ECJ that used apparently minor rulings as opportunities to promote broad and powerful supranational principles.[6]

Significant formal institutional change did not occur, however, until the mid-eighties with the adoption of the Single European Act (SEA) in 1986.[7] The SEA was a package that combined the creation of a Single European Market (SEM)—a fully integrated economic area in which people, capital, goods, and services would move freely—and decisionmaking reforms designed to secure the implementation of the SEM. Two important features of these reforms were the introduction of qualified majority voting in the Council for SEM matters, and a new legislative procedure (the cooperation procedure) with a second reading by the Council after the Parliament's opinion. The two reforms reduced power asymmetry in the Community polity and increased the role of Parliament as an agenda setter vis-à-vis both the Council and the Commission.

The 1991 Treaty of the European Union (TEU)—better known as the Maastricht Treaty—added a third main legislative process (the co-decision

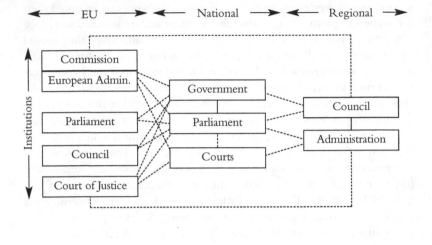

Figure 4.1   The EU Nonmarket Environment: Evolutionary Complexity

procedure), which increased the involvement of the Parliament in law-making.[8] In addition, the TEU gave constitutional recognition to regions, and laid the foundation for the Economic and Monetary Union (EMU) in Europe. The institutional implication of EMU is the creation of the European Central Bank (ECB), an independent supranational body in charge of monetary policy for a subset of EU members.

Let us put this brief description of the evolution of the institutional structure of the EU into the larger perspective of the nonmarket environment of European firms. Figure 4.1 helps understand why I consider this environment to be characterized by "evolutionary complexity." Its complexity comes from the intermingling of the influences of three different levels of governance—EU, national, and regional—each with different actors. Evolutionary complexity reflects the variation of hierarchy and prerogatives along two dimensions. First, there have been shifts in the relative importance of the three levels of governance. Second, at each level, relationships between actors have changed. From an initial de facto situation where the national level was prominent and the supranational was largely confined to the Commission with respect to policymaking, the European nonmarket environment of firms has evolved toward a truly multilevel context with elaborate rules of power sharing among actors.

III. A Decision Map for Lobbying The European Union

In the previous section, I described a polity that is complex with multiple access points for firms seeking to influence decisionmaking. The focus of this section is to determine patterns of strategic preferences of firms engaging in lobbying activities in that complex polity. I first consider issues related to choosing and approaching a target for lobbying, and then synthesize these strategies into an integrated decision map.

Choice of a Target

Among the four institutional cornerstones of the European Union, three of them—the Council, Commission, and Parliament—are key targets for lobbying policymaking.[9] To discuss how one chooses among these three targets, I briefly describe a standard pattern of policymaking in the Union.[10] I use as a benchmark the simplest procedure, or the so-called consultation procedure, followed by a discussion of the more complex cooperation and co-decision procedures.[11]

As depicted in figure 4.2, initiation of the policymaking process starts with an impetus from the Commission, the Council, or the Parliament. Whereas the Commission is the sole body to table a new proposal, the Council has both the political and constitutional (Article 152) powers to induce the Commission to act. And, since the adoption of the Treaty of the European Union, the Parliament has the constitutional right (Article 138b) to request the Commission to submit proposals. Following this, the Commission drafts a proposal. Then, one or several of the twenty-three Directorates-General work in close contact with a host of advisory and consultative committees, comprised of national public officials and private representatives. This work is carried out under the scrutiny of the Commissioners' cabinets.

Once accepted at the level of the Commission, the newly elaborated text goes to the Council. At this stage a working committee examines the text and submits it for evaluation to the European Parliament. Also, where appropriate, the text is referred to the Economic and Social Committee, and, since Maastricht, the Regional Committee.[12] Whereas the latter two

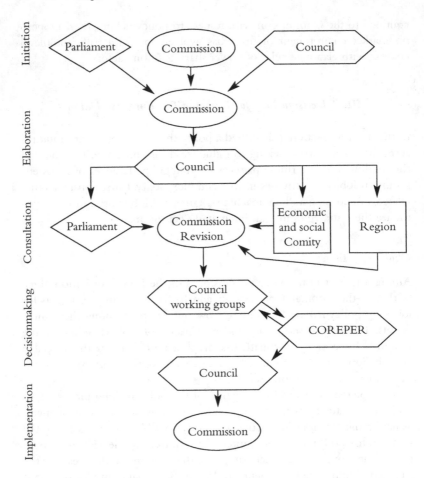

Figure 4.2   A Synoptic Representation of the Simplest Policy Making Process (Consultation Procedure) in the European Union

still play a minor role, the weight of the Parliament has been growing and its opinion is carefully considered.

Once the working party has accomplished a detailed first reading and the advisory committees have given their responses, the Committee of Permanent Representatives (COREPER) examines the proposal. It usually considers the general orientation of a proposal as well as its political implications.[13] In the case of consensus within the COREPER, the proposal is quickly ratified by one of the specialized Councils. In case of disagreement, the matter has to be negotiated by the related Council, or by the higher level of the Council of Foreign Affairs, or even by the European Council (heads of state). If disagreement remains at these higher levels, the proposal can either be sent back to the Commission with requests for change, or referred to a future meeting in the hope of resolving differences. Accepted proposals are then implemented by the Commission, if they are directly binding, or by national governments if they are indirectly binding.

Our discussion so far indicates that a major factor driving the choice of a target is the stage of the policy process for a given issue. Very early in the process, the Commission tends to be the most appropriate target, whereas later intervention should rather focus on the Parliament or on the Council. Access to decisionmaking is generally easier early in the process due to the generalization (and politicization) of discussions at the level of the Council. The extent of existing EU involvement in a policy domain affects its relative likelihood of intervening at any of the various policymaking stages. In particular, implementation grows in importance with the development of the EU body of law.

Looking at the policy domain also influences the choice of the target in relation to the type of policy process. The power and responsibilities of the different targets varies significantly across domains. In particular, the powers of both the Commission and the Parliament are greater for domains where the cooperation or co-decision procedures apply.[14] The cooperation procedure requires that the Council of Ministers can only unanimously alter those amendments by the legislature that the Commission supports. The proposal of the Parliament becomes legislation if it is accepted by a qualified majority of the Council of Ministers. The co-decision procedure gives the Parliament a right to veto legislation independently from the Commission and without a possibility of overrule by the Council even at unanimity voting.

In addition to these factors linked to the issue, there are factors linked to the target. It is crucial for any potential lobbyist to remember that the European Union remains an evolutionary polity. Existing actors strategically engage in complex interactions to further selected interests and agendas. In particular, the supranational actors, the Parliament and the Commission, are always looking for ways to increase their influence, and from this perspective, tend to welcome the help of private actors. In contrast, the Council may often convince particular interests of the need to reach political compromises. From this perspective, choosing the Council as a target requires a careful assessment of sophisticated voting patterns, taking into consideration not only the number of votes of individual countries but also their preference orientation vis-à-vis other members.[15]

To conclude on this first decision node, potential lobbyists may well choose more than one target. This might be particularly appropriate if one wishes to get things started. But then it becomes difficult to closely monitor the follow-up of pressure between targets that might be competing against each other.

Strategic Choices in Approaching Target(s)

Once a target is chosen, the lobbyist has to come up with a strategy to approach it. We focus on two generic strategic dimensions. First, the lobbyist must choose the route, that is, whether to use collective platforms of activities (indirect route) or not (direct route). Second, the lobbyist must determine how to organize action regarding both the geographical and the functional dimensions. Collective platforms can either be organized at the national level or at the European level, and for each of these levels, follow either a sectoral or cross-sectoral logic. Further, one must distinguish between European platforms that are associations of national platforms and European platforms that include firms' individual membership. Choosing a direct route minimizes the issue of organizational mode but does not eliminate it. To increase their ability to influence the target, firms that choose not to use institutionalized platforms may coalesce through ad hoc coalitions.[16]

A priori, none of the options resulting from the combination of choices of route and organizational mode is consistently better than the other ones. Yet, in practice, firms tend to favor the use of national platforms.[17] This is due to several factors. First, access through national platforms avoids the difficulty of reaching compromises with interests in other member states. Second, the political influence of firms is often greater at home than at the international level. Third, national platforms have had a longer experience of lobbying, and thus there are positive network externalities. Fourth, the national platform is the privileged way to access the Council and its working groups. Fifth, national platforms are useful both at the national and European levels (multiple use).

However, there are two big drawbacks with the national path. The first is that support from a single government becomes insufficient with the introduction of qualified majority voting in an increasing number of domains—in particular all issues related to the internal market.[18] Second, even if support from a single government might still be sufficient, integrative bargaining between member states requires the use of tradeoffs and complex issue-linkage, which might push the single government at hand to abandon these specific interests in favor of a general agreement.

In contrast to national platforms, the use of European platforms with collective (or a mix of collective and individual) membership has traditionally been considered as a poor choice, primarily because it is hard to come up with common denominators among a large number of firms and delegations. In addition, large peak associations suffer significantly from free riding behavior.[19] However, recent evidence has begun to show that transnational platforms can achieve important results.[20] There is no doubt that the explicit willingness of the Commission to establish regular contacts with Euro-groups is a strong incentive for using the transnational route, but we will later discuss other facilitating factors. Finally, direct lobbying has the absolute advantage of either avoiding compromises (if a firm is acting individually) or making them unproblematic through the tailoring of ad hoc coalitions. The opposite side of the coin is that visibility of action tends to remain lower than with the other routes.

Our discussion so far has highlighted the general advantages and drawbacks of different strategic options. Individual firms do choose, however, on specific advantages and drawbacks in conjunction with their choice of a target. The latter does constrain the array of strategies but leaves room for the influence of other factors. We briefly consider here two sets of factors that shape these specific concerns. The first comprises individual characteristics of firms, mostly their resources, and their domestic positions. The second includes characteristics of the sector of activity of these firms.

First, a firm's level of available resources affects its preferences. Large multinational enterprises tend to place great value on freedom of maneuver and discount the costs associated with maintaining the freedom. They therefore tend to overstate the drawbacks of "orthodox" European platforms—that is, platforms that are federations of national associations—and favor either a direct route or an indirect one with direct membership in the mediating group. In the latter category, the European Round Table of Industrialists (ERT) is a cross-sectoral example, whereas the Association of Petrochemicals Producers (APPE) is a sector-specific example.[21] In contrast, small and medium enterprises favor joint over individual action and collective membership as a possible means to dilute asymmetry with larger firms.

Another important element pertains to the domestic importance of a firm. Domestic importance may come from a particular market position, such as state monopoly or market dominance, or from larger societal considerations (e.g., the importance of farmers). Further, governments might consider some firms to be "flagships" and thus be particularly receptive to their demands. In sum, although domestic importance often covaries with domestic market shares, it is not simply a reflection of size.

Second, firms include in their strategic analysis contextual factors originating from their sector of activity. In particular, a sector's characteristics include the degree of concentration of activity and the level of competition within it; the structural variance of firms; the maturity variance of products; the politico-economic structures of governance within the sector; and the intensity and "commonality" of threats. Generally speaking, firms are more likely to choose a sectoral mode of organization when the variance of firms' structures and of products' maturity is low, when there is an acute sense of common threat, and when there has been past practice of market control.

The Integrated Framework

On the basis of the factors discussed above, we can now develop a synthetic account of firms' preferred lobbying strategic bundles, including choices of target, route, and organizational mode. Table 4.1 offers a decision map for firms that embark upon lobbying at the EU level.

We derive firms' strategic preferences (column 5 in table 4.1) from the values of four dichotomous variables considered sequentially. The variables are: (a) EU structure of authority over existing legal framework (column 1); (b) resources of the lobbying firm (column 2); (c) national political potency of the lobbying firm (column 3); and (d) cohesion of the sector of activity of the lobbying firm (column 4).

Table 4.1 runs counter to the tendency to advocate the use of a multiple access strategy. While firms do benefit from multiplying the ways in which they lobby the EU, it seems clear that some of these strategies are better than others.[22] Following the logic of the "distributive politics spreadsheet," which is presented in the introduction to this volume, the choice of a lobbying strategy depends both upon the magnitude of the demand that it generates (demand side) and upon its potential effectiveness (supply side). The basic dilemma for firms is that they would prefer not to have their demands diluted through aggregation with other firms or actors but at the same time know that going alone is both costly and on average less likely to generate political action. The array of strategies in table 4.1 reflects a varying assessment of this basic tradeoff.

Table 4.1 A Simplified Array of EU Lobbying Strategies (outside of EMU*)

EU Structure of Authority over Existing Legal Framework	Firm's Resources	Firm's National Political Potency	Structural Cohesion		Lobbying Strategy (target; route; organizational form)
Separate (COM or COUN)	Large	High	Low	L1	COM (nat. ties) or COUN; direct using national connections
			High	L2	COM (nat. ties) or COUN; direct or indirect through Euro sectoral platform
		Low	Low	L3	COM (general) or COUN; direct or indirect through Euro cross-sectoral platform
			High	L4	COM (general) or COUN; direct or indirect through Euro sectoral platform
	Small	High	Low	L5	COM (nat. ties) or COUN; indirect through national/regional cross-sectoral platform
			High	L6	COM (nat. ties) or COUN; indirect through national/regional sectoral platform
		Low	Low	L7	COM (general) or COUN; indirect through Euro federated cross-sectoral platform
			High	L8	COM (general) or COUN; indirect through Euro federated cross-sectoral platform
Shared (COM and COUN)	Large	High	Low	L9	COM/COUN; direct using national connections
			High	L10	COM/COUN; direct or indirect through Euro sectoral platform
		Low	Low	L11	COM/COUN; direct or indirect through Euro cross-sectoral platform
			High	L12	COM/COUN; direct or indirect through Euro sectoral platform
	Small	High	Low	L13	COM/COUN; indirect through national/regional cross-sectoral platform
			High	L14	COM/COUN; indirect through national/regional sectoral platform
		Low	Low	L15	COM/COUN; indirect through Euro federated cross-sectoral platform
			High	L16	COM/COUN; indirect through Euro federated sectoral platform

*Considering the subgroup of EMU members would force us to consider the European Central Bank as a potential lobbying target. In table 4.1, this would affect strategies L1 to L16. Given the high independence of the Governing Board, the target is more likely to be the Council of Governors that includes the governors of the national central banks. Accordingly the national indirect approach is likely to be favored.

The four variables that we use to derive the array of strategies relate to the three dimensions of lobbying choices—target, route, and organizational form—discussed above. The EU structure of authority (column 1) relates to the choice of a target, the resources of the firm (column 2) to the route, the cohesion of the sector (column 4) to the organizational form, and the national political potency (column 3) to the choices of a target and the organizational form.

Before turning to an in-depth discussion of the implications of table 4.1, it is necessary to discuss a few definitional issues as well as some basic simplifying assumptions. First, regarding the definition of variables, the structure of authority and the size of the resources are straightforward, but the other two require specification. The national political potency of a firm reflects the domestic political clout of a firm weighted by the size of the domestic country. Accordingly, one would assign a low value to the national political potency of a firm with a high degree of clout in a small country such as Luxembourg. Turning to the cohesion of the sector the definition of sectoral cohesion includes factors such as the concentration of a sector, the relative convergence of firms' structures, and common threat perceptions. One complicating aspect of sectoral cohesion is that firms with different resources tend to have different "horizons": those with large resources perceive sector cohesion as the European playing field, whereas firms with limited resources see it as the national playing field.

Our decision map rests upon some key simplifying assumptions. First, we use dichotomous variables. Second, table 4.1 depicts strategic options for domains with existing supranational authority. Third, we limit our attention to two generic EU targets: the Commission and the Council.[23]

Table 4.1 suggests a decision path. First, firms consider the structure of EU authority over the issue of lobbying. Then, the size of their resources helps them to reduce their range of options. Lastly, they take into account both their national political potencies and the contexts of their sector of activity.

Regarding the choice of a target, the more extensive the EU legal framework, the more likely firms will lobby at the implementation stage. At that stage, the EU structure of authority is separate from the Commission as the main actor. Overall, however, firms will face a situation of shared competence. In domains such as industrial policy, environmental policy, transport policy, and trade policy, the Commission shares authority with the Council. Alternatively, the Council might remain the key target in domains with significant EU involvement, as in the case of energy, foreign policy, or exchange rate policy in the newly formed European Monetary Union.

The influence of resources over strategic options is significant in table 4.1. Firms with abundant resources can afford to go either directly or indirectly to the target, whereas small firms must approach the target indirectly. In addition, large firms can also push for Euro-platforms with direct membership, whereas small firms normally use Euro-federated platforms—that is, super-federations of national associations.

High levels of national political potency tend to push firms to use national channels of action. This means that they favor, as a target, the Council over the Commission, and inside the Commission, deal with the Commissioners' Cabinets or resort to national networks within the administration. Large firms can effectively use these national channels on their own, while smaller firms must use national collective platforms. In contrast, firms with low national political potency are more likely to target the Commission in a general way, and when they target the Council, they focus more on technical working committees.

A high degree of sectoral cohesion is decisive for the choice of whether a firm pursues a sectoral mode or an effort to organize interests. This strong effect is due to the fact that when a sector enjoys high cohesion, joint demands at the sectoral level do not dilute individual demands; at the same time, the costs of organizing political action remain low. These two positive factors help compensate for the potentially smaller impact of sectoral action as compared to cross-sectoral political action. Also, they can entice large firms with high national political potency (L2, L10). High sectoral cohesion does not make large firms systematically prefer indirect sectoral action to direct lobbying; rather, it only determines the mode of organization if large firms choose an indirect route. Conversely, when a sector lacks cohesion, joint demands at the sectoral level dilute individual demands as much as joint cross-sectoral demands, while also being comparatively much weaker in terms of the capacity to generate political action. In cases of low cohesion, the drawbacks of sectoral action are such that they eliminate strategic indeterminacy for large firms with high national political potency (L1 and L9).

The choice of constraining variables helps produce relatively precise, if not fully determined, preferred lobbying strategies. Strategies L1 to L8 refer to one target (either the Commission or the Council), one organizational mode, and one or two possible routes. Strategies L9 to L16 refer to two targets, but with a preference for one, and one or two possible routes. The size of firms' resources is the basis for indeterminacy. Large resources enable firms to always consider direct action irrespective of the degree of sectoral cohesion and of their national political potency.

Our decision map links a firm's individual situation with its most preferred strategic bundle. A topic for further interest is to compare the political effectiveness of different strategic options. Coming up with a comprehensive ranking of L1 to L16 does not make much sense given the contextual differences, but an examination of the relative effectiveness of subsets of options helps elucidate the issue of how firms can improve their bargaining position. For instance, one can consider the case of firms with small resources when EU competency is shared between the Council and the Commission (options L13 to L16). One can argue that L14 yields on average the highest return to small firms. Individual demands are the least likely to be diluted through joint action, the costs of organizing action remain moderate, and, given the high visibility at home, collective action is very likely to affect the EU polity. By contrast, strategy L15 yields the lowest return. Individual demands are diluted twice, once through national aggregation across different sectors of activity and then through the aggregation of the demands of national associations. The costs of organizing are very large, which lowers the prospects for delivering a message that can have any significant impact. Following a similar logic, L16 yields a higher return than L13. The upshot of this discussion is that small firms that wish to improve the return of their nonmarket strategies should first try to improve the cohesion of their sector of activity, and then try to increase their national political potency.[24] None of these two efforts to change the game are easy. Increasing national political potency is impossible for firms located in small countries. However, there are feasible ways to promote change. For instance, greater sectoral cohesion can be achieved through an increased specialization of sectors of activity, leading to narrower sectors that are more likely to become cohesive. Collective action is then easier to organize and demands will be less diluted. Specialization might also increase national political potency, as can efforts to develop action more geographically concentrated at the national level. The increasing relevance of regions in most EU countries opens up avenues of increased influence for small firms.

Turning to large firms, ranking effectiveness inside the subset L9 to L12, for instance, is problematic due to the indeterminacy of individual preferences. Further, even when controlling for this indeterminacy and trying to rank only options with an indirect route to the target, determining a specific ranking is difficult. The drawbacks of cross-sectoral action are much fewer than for small firms, given the limited number of firms engaged in such action, and potentially strengthen the volume of the message delivered to the EU polity. Accordingly, large firms might not want to invest in increasing the cohesion of their sector of activity. Also, a higher national political potency might not be relevant for large multinational firms. Lobbying the EU polity outside the national channels might, but need not, be more effective.

A further issue of discussion is the relative need for involvement in lobbying at the EU level. Can one a priori say that in some cases there is little gain to be expected from specific EU behavior and thus that firms with national ambitions are the only ones who should lobby at the national level? First of all, table 4.1 applies to cases where EU competence is significant. When there is little competence, giving priority to the national level is much more appealing, especially for small firms that can pretend to help institutional EU actors air new ideas in elite circles. When EU policy involvement is significant, table 4.1 suggests that some firms can avoid choosing between national and supranational lobbying and can thus do both with little loss of effectiveness. This applies best to small firms with high national political potency when there is either predominance of the Council or shared competence between the Council and the Commission (L5, L6 with Council competence and L13, L14). These firms organize nationally to act upon their government and thus they need not be concerned about the EU level per se. Hence, for them, the national level of activity is of multiple use.[25] Specific EU action is more useful when the target is the Commission (L5, L6 with Commission competence). National ties inside the Commission might not coincide with the prevailing home government, and thus there might not be an automatic transfer from the national to the supranational level. Due to non-overlapping election cycles and reelection options, Commissioners from one country often do not share the same party orientation, and thus might disagree over the domestic political potency of firms. Somehow, paradoxically, large firms can less afford not to choose between national and supranational action. They do not have to organize at the national level, especially if they have a high political potency, and thus lobbying the EU polity implies more tailored use of national channels.

However, one should consider a few words of caution regarding the use of the decision map depicted in table 4.1. First, strategic options L1 to L16 reflect the most preferred *individual* options for firms. When firms act jointly through collective platforms, they may not all be pursuing their most preferred option. A good example is the European Round Table of Industrialists (ERT). This cross-sectoral platform corresponded to the most preferred option of large firms with low national political potency and low sectoral cohesion. Nestlé and Philips nicely fit in this category. However, according to table 4.1, it was not the most preferred option of large firms with high national political potency. How can one then explain that Renault, for instance, joined the ERT? There was obviously bandwagoning behavior that occurred once the core structure came into concrete existence. The framework cannot account for such considerations, but it does help explain why some actors, like Nestlé and Philips, were the most active in setting up the ERT. The second reason for caution concerns the potential importance of the interdependence between the target and the one who seeks to influence. Active efforts by targets to become targets might distort the preferences in table 4.1. Still, the theoretical synthesis provides a useful benchmark to calibrate lobbying strategies.

IV. An Application to European Firms
Concerned with Opening Up Asian Markets

We now turn to a brief application of our decision map to examine the case of European firms wishing to penetrate Asian markets. We proceed along two lines of inquiry: the policy domain (mostly trade and investment policies) and selected sectors of activity. Regarding the latter, we refer to the sectors covered in part three of this volume—automobile, aerospace, financial services, and the software industry—and to other sectors considered to offer key opportunities to European firms in Asia, namely chemicals, infrastructure, energy, food business, and luxury goods.[26]

Policy Domain, Sector of Activity:
EU Competence and Structure of Authority

Asian markets offer large opportunities for both European investment and exports—particularly if one notes that European firms have been behind both Japanese and U.S. firms on both issues.[27] The extent of EU involvement in these two external economic issues has been significant. This is especially true for trade policy, which is common to all members and relies heavily on legal regulation. Involvement in investment issues has emerged more recently, either through negotiations over trade-related investment measures in the General Agreement on Tariffs and Trade (GATT)/World Trade Organization (WTO), or through the work of the Organization for Economic Cooperation and Development (OECD) on the failed effort to develop a multilateral investment regime.

In terms of authority, the Commission and the Council are the key players in matters of external economic relations. The Treaty of Rome gives broad powers to the Council in the domain of trade policy.[28] According to Article 113, the Council issues negotiating directives for the Commission, which conducts negotiations for the Community. The Council also appoints a special committee, the so-called 113 Committee, that assists the Commission. It has the power to endorse any major trade agreement with a qualified majority (Article 113[4] and 114 EEC).[29] The Commission may intervene in the discussions inside the Council, and also in the Committee of Permanent Representatives (COREPER). The European Parliament has had few prerogatives, except for the power of assent conferred by Article 228 when an international agreement has "important budgetary implications for the Community." [30]

The EU's authority over foreign investment is less explicit. When investment is linked to trade, the above structure applies. However, when discussions take place in a more general context, the EU's involvement is less codified, and the practice has been to reproduce the shared system of authority between the Commission, acting as the agent, and the Council, acting as the principal.

When lobbying the Commission for trade and investment issues in Asia, firms have several possible specific targets. First, on a cross-sectoral basis, firms can approach the Commissioner for External Relations and the Directorate-General for External Economic Relations (DGI with its two subdivisions DGIA and DGIB). The Commissioner for External Relations represents the EU in trade negotiations, both multilateral and bilateral, and in several other international economic fora. In addition, the Commissioner is involved in missions to promote EU interests in line with a general "Market Access Strategy for the European Union" that reflects "a more proactive stance in tune with the real needs of European exporters in an increasingly interdependent global economy."[31] In terms of specific Asian actions, DGI partly sets the agenda of Asia-Europe Meetings (ASEM) between heads of state and government. DGI is also responsible for the follow-up of these meetings, including the implementation of the Investment Promotion Action Plan (IPAP), the Trade Facilitation Action Plan (TFAP), and the ASEM Business Forum. The Commission also has developed the Asia-Invest program managed by DGIB. The program, launched in 1997 for an initial five-year period, aims to promote two-way trade and investment between Asia and Europe, especially for small and medium enterprises (SMEs).[32]

Besides DGI, firms can also generally target the Directorate for Internal Market and Industrial Affairs (DGIII), the Directorate-General for

Science, Research, and Development (XII), and the Directorate-General for Enterprise Policy, Distributive Trades, Tourism, and Cooperatives (DGXXIII). DGIII has been very active in promoting global competitiveness, especially in the capital goods industries (of special interest for firms in aerospace and infrastructure). It also managed the European Strategic Program for Research and Development in Information Technology (ESPRIT) research program until October 1998, and runs the EU-Japan Center for Industrial Cooperation. Most firms—but especially those in the aerospace, energy, telecoms, and software industry—can benefit from the action of the Directorate-General for Science, Research and Development (DGXII). The SEA codified the Research and Technological Development policy (R&TD), and the TEU reinforced it, notably with the objective of encouraging the international competitiveness of European firms (Art. 130f, EC). There has been a large set of R&D framework programs, especially in telecommunications and information technologies, that seek to help European firms become more competitive on the Asian markets.[33] DGXXIII is of particular interest for SMEs for its activities of information distribution and guidance in world markets. Notably, it runs the Business Cooperation Network (BC-Net) and the Bureau de Rapprochement des Entreprises (BRE), both of which now have a "Support for Asia" connection inside the Asia-Invest program.

Additional incentives to target the Commission might come from sector-specific interests. Efforts by various DGs to promote a more integrated market in Europe might spill over into the external capacity of firms, which in turn will bring domestic benefits. From this perspective, firms in telecoms and information technology sectors might target DGXIII (Telecommunications, Information Technology, and Industries), firms in energy and infrastructure might target DGXVII (Energy), and firms in financial services might target DGXV (Financial Institutions and Company Law).

In sum, with respect to the choice of the target, there is significant EU involvement in trade and investment issues, with both the Council and the Commission being active. Although the Council keeps the upper hand, it has allowed and encouraged the Commission to launch a host of initiatives that make this body a very attractive target.

Firms' Individual and Collective Characteristics

A perusal of firm characteristics in certain key sectors provides a hint as to broader sectoral trends.

- The automobile sector includes primarily large players (Fiat, Volkswagen, Renault, Peugeot, Daimler-Chrysler, BMW, General Motors Europe, Ford Europe). Several of these firms have high national political potency, as they both come from large countries and have had strong ties with their governments. Also, this sector lacks cohesion due to a high level of variance in the structures of firms and a growing perception that the Japanese threat of the 1980s has vanished.
- The chemical sector comprises large multinational firms (Novartis, ICI, Rhône-Poulenc, Hoffman–La Roche, Bayer) with a mixed (but declining) record in terms of national political potency. But the sector has high cohesion, not least because of established collusive practices of market control.
- The aerospace industry has been the domain par excellence of a few large firms—British Aerospace (BAe), Aerospatiale, and Daimler-Benz—with high national political potency. Due to the limited number of firms that count in this field, sector cohesion is not really an issue.
- In the sector of infrastructure (transport, power industry, construction), cohesion is low. SMEs coexist with large groups (ABB, Iberdrola, Saint-Gobain, Lafarge, Bouyges, Thiessen, Altshom, Siemens, Krupp, Vivendi) that have a mixed valuation of national political potency.
- National political heavyweights still very much characterize the energy sector (oil/gas extraction and distribution), which is dominated by a few large firms (BP, Shell, Elf, Total, Petrofina). The sector enjoys a relatively high level of cohesion in terms of firms' structure and of past collective practices.
- In the telecom manufacturing/equipment industry, a few large multinational actors with modest national political potency (Siemens, Alcatel, Nokia, Ericsson) coexist with smaller firms, which results in a low sectoral cohesion. Telecommunications service is still highly influenced by the monopolistic structures (large players with special domestic position) that have characterized it for decades. As a result, there is hardly any sectoral cohesion, and still high national political potency.
- The software industry consists mostly of SMEs with low national political potency.
- In financial services, banking is one of the continent's more fragmented industries. There is high variation in size (from big commercial banks like HSBC holdings, Lloyds TSB, ING group, UBS, and Credit Suisse to small local savings banks) and a large number

of actors in every category. In addition, there is still a great deal of political pressure to create national champions, in particular in countries like France and Germany.

- The situation is similar in the domain of insurance, even though a process of pan-European consolidation has gradually emerged.
- Finally, in the food business and luxury goods, large companies (Unilever, Danone, Nestlé, and to a less extent LVMH, Gucci, and Hermès in the luxury goods sector) coexist with many smaller firms. High sectoral variance and low national political potency characterize the sector.

Lobbying Strategies

The above discussion of the four variables considered in the framework indicates that EU lobbying strategies for firms wishing to penetrate Asian markets cluster around strategies L9 to L16 in table 4.1. Large firms that have high national political potency will mostly follow L9 but tend to combine it with L10 when there is some concentration of resources. Such combinations should be used by firms in energy, automobiles, and to some extent in the aerospace industry. In addition to national connections, automobile makers can resort to the Association of European Automobile Constructors (ACEA) and oil companies possess the platform of the Association of Petrochemicals Producers (APPE). In the aerospace industry, collective action would at best go through transnational cooperation. The Groupement d'Intérêt Economique (GIE) structure used by Airbus is an example of such a sophisticated type of transnational organization. In contrast, large firms in the more fragmented sectors of financial services and infrastructure can be expected to focus exclusively on national connections (L9).

Large firms that have low national political potency will follow L11 and L12. The ERT corresponds to the indirect route in strategy L11. Large players in the food industry (Nestlé, Danone, Unilever) and telecoms equipment (Nokia, Siemens, Ericsson) have been extremely active inside the ERT. Large chemical firms are likely to rely more heavily on the sectoral approach (L10), and favor action through the European Chemical Industry Council (CEFIC), a Euro platform with a mix of direct and indirect membership.

Smaller firms, most of which possess low national political potency, can mostly be expected to pursue L16, that is, an indirect route to the Commission through federated Europlatforms. These platforms may be multiple and represent different subsectors. In the telecom sector, for instance, firms can choose among three European confederations—the Associations of Manufacturers of Insulated Wires and Cables (EUROPACABLE), the European Association of Manufacturers of Business Machines and Information Industry (EUROBIT), the European Conference of Associations of Telecommunications (ECTEL).[34] Cross-sectoral modes of organization (L15) are less attractive for small firms, except through the overarching body of the Union of Industrial and Employers' Confederation (UNICE).

V. Conclusion

Building on existing conceptual work, this chapter developed a comprehensive analytical framework to trace firms' EU lobbying strategies. I considered choices of targets, route, and organizational modes, leading to an array of sixteen types of strategies derived from four key variables. These include the EU structure of authority over the issue, the size of a firm's resources, its domestic position, and the cohesion of the given sector of activity.

The framework helps to reduce the complexity of the decisions firms face when they wish to engage in EU lobbying and provides them with general strategies. I also examined how this applies to the cases of European firms wishing to penetrate Asian markets, looking at various sectors of activity—automobile, aerospace, chemicals, energy, financial services, food and luxury goods, infrastructure, the software industry, and telecommunications.

Future work should relax some of the limiting assumptions used in this chapter. In particular, the framework should be extended to reflect better the intricacy of strategic interdependence between targets and lobbying firms. This is a daunting task, however, because the additional richness should not lead to general indeterminacy, as is often the case with the analysis of complex strategic interactions.

Notes

1. According to Sargent (1993), firms have used lobbying to seek the following objectives: timely compliance with EU legislation; damage limitation; safeguarding the general trading environment; insurance policy; exploitation of new business opportunities and ideas.
2. For examples of both types of work, see in particular Greenwood, Grote, and Ronit (1992b), Mazey and Richardson (1993), Pedler and Van Schendelen (1994).

3. Greenwood, Grote, and Ronit (1992a).

4. Streeck and Schmitter (1991), Lanzalaco (1992), Traxler and Schmitter (1995).

5. See chapter 1 by Aggarwal in this volume. The paper does not consider tactical considerations that pertain to the methods and techniques of lobbying. Methods are either direct (consultation, hearings, interviews, phone conversations, letters, personal visits) or indirect (reports, studies, use of brokers and consultants). Techniques include issue manipulation, clientelism, argumentation, coalition building, agenda formation, rewards and compliance, and public relation campaigns.

6. For a discussion on the increasing role of the ECJ, see for instance Alter (1998) and Mattli and Slaughter (1998).

7. For details on the Single European Act, see for instance Armstrong and Bulmer (1998) and Wallace and Young (1996).

8. It is unclear, however, how much additional power this gave to the Parliament. See Garrett and Tsebelis (1996).

9. While a powerful actor, the Court of Justice intervenes only at the adjudication stage.

10. Detailed discussion of policy processes in the European Union can be found in several books, including Andersen and Eliassen (1993), Keohane and Hoffman (1991), Lodge (1994), Nugent (1994), and Sbragia (1992).

11. The consultation procedure can be found in the Treaty of Rome. The cooperation procedure was introduced with the Single European Act, and the co-decision procedure with the Treaty of the European Union—the Maastricht Treaty.

12. The Economic and Social Committee (ESC or ECOSOC) is a tripartite body with representatives from producer, labor, and consumer interests. Delegates are appointed by national governments rather than by national or European organizations.

13. They may send the proposal back to a working group for further detailed consideration.

14. The cooperation procedure applies to domains such as common transport policy, European Social Fund, and the European Regional Development Fund. The co-decision applies for instance to the free movement of workers, culture, consumer protection, mutual recognition of formal qualifications, or harmonization for the purpose of completing the internal market. For a detailed list of the policy domains covered by the various procedures, see Nugent (1994), pp. 318–322.

15. Lobbying a small EU country might produce better results than lobbying a larger country when the latter is a clear outlier in terms of preferences over an issue. See Garrett and Tsebelis (1996).

16. Pijnenburg (1998) documents the cases of two prominent ad hoc coalitions, the European Committee for Interoperable Systems (ECIS), and the Software Action Group for Europe (SAGE), in the software industry.

17. See, for example, Coen (1997).
18. Analysts have highlighted the importance of the Single European Act to account for a change in lobbying activities. See for instance Coen (1997) and Cowles (1997).
19. See Averyt (1977), as well as contributions in Mazey and Richardson (1993) and in Greenwood, Grote, and Ronit (1992b).
20. See, for instance, Greenwood and Ronit (1992), who focus on the pharmaceutical industry.
21. Direct membership may not always produce good results, as can be seen from the case of the Committee of European Community Automobile Makers (CCMC). See McLaughlin and Jordan (1993) on p. 132.
22. See Bennett (1997) for a similar position.
23. Relaxing these assumptions would prevent a synthetic account without bringing much additional insight on firms' basic preferences. Considering cases where the existing EU competence is absent or insignificant would enlarge the set of targets to the European Parliament. Indeed, when EU authority is absent, firms seek to lobby at the early stages of lawmaking. As we discussed above, the Commission always drafts new legislation and the Council formally makes law. But, under the cooperation procedure and the co-decision procedures, the Parliament can also become a valuable target of influence given its conditional agenda setting power.
24. Of course, they should also try to increase their resources so that they become large firms. This is, however, out of the reach (and even not on the drawing board) of most of small firms.
25. See Bennett (1997) and Coen (1997) for empirical evidence on this point.
26. EIJS (1997).
27. European Commission (1996).
28. Given the absence of a separate Council of Ministers for trade, discussions on trade policy are usually addressed by the Council of foreign ministers. Hayes (1993), p. 37.
29. In the initial stages (that is adoption of negotiations' directives proposed by the Commission, or subsequent directives' amendment), the EEC Treaty mentions that the agreements must be accepted by the Council at unanimity rule (Article 114 EEC).
30. The European Parliament is notified about agreements, with some discussion of trade matters taking place inside the Parliament's External Economic Relations Committee. Nugent (1994) p. 390.
31. European Commission (1996).
32. For more information, see http://www.asia-invest.com.
33. Among the best known R&TD framework, one finds Basic Research in Industrial Technologies for Europe (BRITE), ESPRIT, and Research and Development in Advanced Communications Technologies for Europe (RACE).
34. For a detailed study on these organizations, see Schneider et al. (1994).

References

Alter, Karen J. (1998). "Who Are the Masters of the Treaty? European Governments and the European Court of Justice." *International Organization* 52 (1), pp. 121–147.

Andersen, Svein S. and Kjell A. Eliassen (1993). *Making Policy in Europe.* London: Sage.

Armstrong, Kenneth and Simon Bulmer (1998). *The Governance of the Single European Market.* Manchester: Manchester University Press.

Averyt, William (1977). *Agro Politics in the European Community: Interest Groups and the Common Agricultural Policy.* New York: Praeger.

Bennett, Robert J. (1997). "The Impact of European Economic Integration on Business Associations: The UK Case." *West European Politics* 20 (3), pp. 61–90.

Coen, David (1997). "The Evolution of the Large Firm as a Political Actor in the European Union." *Journal of European Public Policy* 4 (1), pp. 91–108.

Cowles, Maria Green (1997). "The Changing Architecture of Big Business." In *Collective Action in the European Union: Interests and the New Politics of Associability,* edited by Justin Greenwood and Mark Aspinwall. London: Routledge.

EIJS (1997). *Looking Ahead: Challenges, Synergies and Opportunities in Economic and Business Relations between Europe and Asia.* Tokyo: European Institute of Japanese Studies, Stockholm School of Economics.

European Commission (1996). *Investing in Asia's Dynamism. European Union Direct Investment in Asia.* Brussels: EC/UNCAD.

Garrett, Geoffrey and George Tsebelis (1996). "An Institutionalist Critique of Intergovernmentalism." *International Organization* 50 (2), pp. 269–299.

Greenwood, Justin, Juergen R. Grote, and Karsten Ronit (1999a). "Introduction: Organized Interests and the Transnational Dimension." In *Organized Interests and the European Community,* edited by Justin Greenwood, Juergen R. Grote and Karsten Ronit. London: Sage.

Greenwood, Justin, Juergen R. Grote, and Karsten Ronit, eds. (1999b). *Organized Interests and the European Community.* London: Sage.

Greenwood, Justin and Karsten Ronit (1992). "Established and Emergent Sectors: Organized Interests at the European Level in the Pharmaceutical Industry and the New Biotechnologies." In *Organized Interests and the European Community,* edited by Justin Greenwood, Juergen R. Grote, and Karsten Ronit. London: Sage.

Hayes, John P. (1993). *Making Trade Policy in the European Community.* New York: St. Martin's Press.

Keohane, Robert O. and Stanley Hoffmann, eds. (1991). *The New European Community: Decisionmaking and Institutional Change.* Boulder: Westview Press.

Lanzalaco, Luca (1992). "Coping with Heterogeneity: Peak Associations of Business within and across Western European Nations." In *Organized Interests and the European Community,* edited by Justin Greenwood, Juergen R. Grote, and Karsten Ronit. London: Sage.

Lodge, Juliet, ed. (1994). *The European Community and the Challenge of the Future*. London: Pinter.

Mattli, Walter and Anne-Marie Slaughter (1998). "Revisiting the European Court of Justice." *International Organization* 52 (1), pp. 177–209.

Mazey, Sonia and Jeremy J. Richardson, eds. (1993). *Lobbying in the European Community*. Oxford: Oxford University Press.

McLaughlin, Andrew and Grant Jordan (1993). "The Rationality of Lobbying in Europe: Why Are Euro-Groups So Numerous and So Weak? Some Evidence From the Car Industry." In *Lobbying in the European Community*, edited by Sonia Mazey and Jeremy J. Richardson. Oxford: Oxford University Press.

Nugent, Neill (1994). *The Government and Politics of the European Union*. London: Macmillan.

Pedler, R. H. and M. P. C. M. Van Schendelen, eds. (1994). *Lobbying the European Union*. Aldershot: Dartmouth Publishing Company.

Pijnenburg, Bert (1998). "EU Lobbying by Ad Hoc Coalitions: An Exploratory Case Study." *Journal of European Public Policy* 5 (2), pp. 303–321.

Sargent, Jane A. (1993). "The Corporate Benefits of Lobbying: The British Case and its Relevance to the European Community." In *Lobbying in the European Community*, edited by Sonia Mazey and Jeremy J. Richardson. Oxford: Oxford University Press.

Sbragia, Alberta M., ed. (1992). *Euro-Politics. Institutions and Policymaking in the "New" European Community*. Washington, D.C.: Brookings Institution.

Schneider, Volker, Godefry Dang-Nguyen, and Raymund Wurle (1994). "Corporate Actor Networks in European Policy-Making: Harmonizing Telecommunications Policy." *Journal of Common Market Studies* 32 (4), pp. 473–498.

Streeck, Wolfang and Philippe C. Schmitter (1991). "From National Corporatism to Transnational Pluralism: Organized Interests in the Single European Market." *Politics and Society* 2, pp. 133–164.

Traxler, Franz and Philippe C. Schmitter (1995). "The Emerging Euro-Polity and Organized Interests." *European Journal of International Relations* 1 (2), pp. 191–219.

Van Schendelen, M. P. C. M., ed. (1993). *National Public and Private EC Lobbying*. Aldershot: Dartmouth University Press.

Wallace, Helen and Alasdair R. Young (1996). "The Single Market." In *Policy-Making in the European Union*, edited by Helen Wallace and William Wallace. Oxford: Oxford University Press.

Part Three

Case Studies

Chapter 5

From Local to Global:
European Enterprise Software Strategies in Asia

Trevor H. Nakagawa

I. Introduction

Once considered peripheral to the computing world, software has become a central driver of growth in an expanding global economy. Characterized by increasing returns to scale, network externalities, and high fixed costs, first mover advantages are particularly strong in the software industry.[1] With its origins in the U.S. market, it is not surprising that U.S. software firms continue to dictate the pace of innovation while capturing dominant market shares in the fastest growing information technology (IT) region in the world, the Asia-Pacific. Not only have U.S. firms successfully exported their products to Asia, but they continue to gain market shares in several European countries as well. Initially composed of customized solution providers for proprietary hardware designs and specific end-users, European software producers were slow to adapt to the new challenges posed by U.S. firms with standardized software products tied to rapidly innovating computer hardware systems.

Although distinct market and nonmarket conditions led to the creation of the first large, independent, software firms in the United States by the 1980s,[2] it was not clear at the time that European proprietary hardware systems (which usually produced their own software internally) could not compete internationally. Even the world's undisputed mainframe leader of the era, IBM, did not realize how operating system software (i.e., Microsoft) could become the key bottleneck of the evolving microcomputer (PC) market.[3] But since the late 1980s, the widespread demand for the PC and the current Internet e-commerce boom have left few European players positioned to participate in the explosive packaged software market growth throughout Asia.

Despite concerted, decade-long policy efforts by the European Union (EU) and national governments alike to catch up, only a few European software firms are global players today. Because many firms possessed path-dependent competence in firm or industry specific customized software niches, they often sought to become software service providers for their blue chip clients with operations in Asia. While making this organizational shift, rapid innovation in microcomputing processing power in the United States combined with lower prices to create even larger installed bases of PCs for a variety of users throughout Asia. This led to U.S. independent software vendor (ISV) market share dominance in the booming Asian markets. Although European firms have not fared well in the mass-market packaged software market, they have opportunities in the other rapidly growing part of the traded software market, the enterprise solutions segment. Global players like Germany's SAP AG illustrate how prior competencies such as local market familiarity and knowledge of specific users' needs can be leveraged to create standardized, packaged solutions for large multinational users. In an increasingly global market, the software product with the best value-added functionality to any given critical mass of users can compete successfully for market share. Because the European software market remains the second largest in the world, it is not surprising that the few ISV products that have had success in Asia enjoyed dominant market shares at home. Given the significance of reputation and brand in Asian markets, it was also important to establish market share in the United States, the world's largest market. Because the most prominent independent European software firms that exist on the global level are in the enterprise solutions market, this segment will be the primary focus of the case studies in this chapter.[4]

The independent software firm is a relatively new and diverse industry. Today, it has evolved into roughly three major segments: (1) mass-market packaged software, (2) enterprise solutions software, and (3) professional software services.[5] This chapter analyzes the dynamics of the software industry by focusing on traded software for enterprise solutions.[6] Although a large portion of software production is untraded and embedded in electronic equipment or customized for specific clients, independent software product firms continue to be key engines of growth in the economy. Not only are standardized software solutions one of the fastest growing segments in the IT industry, their role in the future evolution of information infrastructures is expected to grow exponentially.[7] In addition, the customized software solutions segment that used to constitute the majority of the European software industry continues to decline in growth as new advances in standardized software capabilities are imported from abroad, particularly the United States. Finally, the focus on this traded software market makes sense for one important practical concern.[8] It is much easier to obtain reliable and comparative data for software products, than try to disaggregate the cost of software in implementing system integration services involving complex manufacturing processes and machinery that often include extensive service related costs (see Appendix 1).

Although the European software market is the second largest in the world, only four European companies ranked among the top fifty producers in the world in 1998—and all serve the enterprise solutions or services market (see table 5.1). Of those, an average of 84 percent of their revenues are from the EU, primarily their own national domestic markets.[9] As an industry that began as a peripheral division of national computer hardware producers, the European software industry was fragmented among competing proprietary hardware designs and differentiated along national and functional divisions. Numerous, small customized solution providers for specific end-users competed among larger telecommunications or electronics giants for various market niches. Such customized software products were not effective exports to a budding Asian mass consumer software market growing at annual rate of over 9 percent between 1990–1995—nearly doubling that of the EU over the same period.[10] Not only are Asian countries attractive as growing emerging markets with healthy economies, but government-industry, high-technology collaboration has contributed to the development of the infrastructure, skilled labor and incentives necessary to help the software industry grow with the rapidly evolving global market.

Thus, as an industry identified as crucial to future economic competitiveness by both EU and Asian governments alike, a focus on the experience of a few successful European software firms in the traded enterprise solutions software segment provides an excellent viewpoint from which to analyze market and nonmarket factors for successful entry into emerging East Asian markets.[11]

This chapter proceeds in four stages. First, I examine the global position of the European software industry. This analysis of the distinct characteristics of an increasingly global software industry dominated by U.S. firms is

Table 5.1 Leading Firms in Traded, Packaged Software (selected firms in U.S.$100,000)

Firm	1984	Rank	1992	Rank	1996	Rank	Country
IBM	3,197	1	11,366	1	12,911	1	U.S.
HP	500	2	N/A	N/A	941	13	U.S.
NEC	300	4	1,840	4	2,263	6	Japan
Fujitsu	200	5	3,535	2	4,754	4	Japan
Microsoft	125	9	2,960	3	8,963	2	U.S.
Olivetti	96	12	708	13	881	14	Italy
Siemens	39	14	1,058	6	1,010	12	Germany
SAP AG	0.183	N/A	6.06	N/A	205	N/A	Germany

Source: Datamation (1985, 1993, and 1997). This includes enterprise solutions revenue. It should also be noted that the gaps in ranking indicate the exit, entry, and/or consolidation of software industry players.

used as a benchmark for understanding the evolution of the European software industry and European government software policies. Second, I examine the market and nonmarket dynamics of key East Asian markets and the position of European firms vis-à-vis their competitors. This forms the context for the analysis of both market and nonmarket strategies of European firms in the third section. The final section is devoted to a presentation of three case studies to provide illustrations of some necessary conditions for European software firm success in Asia. Disproportionate attention will be paid to the case of SAP AG (Systems, Applications, Products), as it is the only major European player recognized as a world leader in any major segment of the software industry.[12] With revenues of over $5.1 billion, employing over 23,700 people in over fifty-one countries, SAP enjoyed the title of being the third largest independent software supplier in the world through 2000.

II. Positional Analysis

Although latecomers to the global market dynamics of the rapidly evolving packaged software industry, some European firms found themselves well positioned to take advantage of their path-dependent, local market advantages. In particular, the most successful European software firms have managed to remain well entrenched in the largest market segment, the B2B (business-to-business) inter-enterprise market. The firms reviewed in the case study portion of this chapter, SAP, Micro Focus (now Merant) and Synon (now CA/Sterling) epitomize the successful transition from a local to global producer of B2B software with three very different results—the first became a global leader, the second became a consolidated entity through acquisition, and the third could not survive the industry shakeout. Having vast experience of local firms' requirements in various industries in the initial phase, these firms were able to successfully leverage their expertise in the operations of large scale enterprises to survive the shift to more distributed forms of computing to produce inter-enterprise solutions, especially for multinational corporations and other large organizations like public utilities. With IT industry growth advancing most rapidly in Asia, the most successful European software firms were the ones able to make strategic alliances with the technology partners and host governments involved in the creation of the new Asian IT infrastructure. By aggressively attacking these large emerging markets, the European software players would have a fighting chance at maintaining the loyalty of their own national clients with multinational operational needs; while simultaneously creating a global, standardized software product of general import to entire industrial segments. However, while Micro Focus barely survived merging with similarly sized firms that were acquired to create Merant Micro Focus in 2000, Synon could not survive being swallowed up by larger market players in buyouts (i.e., first Sterling Software and then Computer Associates to create CA/Sterling). Regardless, they still represent successful attempts at creating packaged software products for an increasingly global market, including making inroads into Asia as part of their core strategy. That the larger, evolving companies that exist at the end of 2000 have incorporated their technologies, products and software solutions into their integrated e-business software packages attests to the strength of their stand-alone successes.

Market Factors

The European Software Industry in the Global Marketplace. As the second largest software market in the world with steady growth at a healthy five percent per annum over the past decade, it is surprising to see that few European software firms have a visible global presence. Because European software producers differentiated along environmental (i.e., proprietary platform), industrial, and national lines, the industry did well at meeting the specialized needs of their clients operating in distinct market segments. In other words, they excelled at producing sophisticated customized software for large organizations, providing continuous service in the process. Client loyalty, proprietary hardware designs, and long-term relationships translated into a competitive dynamic of user responsiveness, low standardization, and re-use rates, specialized knowledge of business processes, and strategic management of technology (see tables 5.2 and 5.3).

But with the rapid proliferation of the inexpensive PC, the Internet, and packaged software, a very different industry dynamic emerged in the United States (see tables 5.4 and 5.5). This is a system based on open (but owned) standards, increasing returns (i.e., high fixed cost, low marginal costs), network externalities, high lock-in and switching costs, and the importance of compatible standards.[13] As a result, European firms have been unable to keep pace with the quality and low price of their U.S. competitors' software products in the packaged mass-market software business. Firms in this $154 billion packaged software industry (1999) have now become household names, such as the $17.4 billion revenues of Microsoft.[14]

Table 5.2 Traded, Packaged Software as a Share of Total Software and Service (%)

Year	1987	1995
U.S.	31.85	36.93
Japan	18.34	23.66
Germany	28.03	39.55
France	22.07	31.75
UK	37.16	42.80
Italy	33.47	35.71

Source: OECD (1997)
Note: "Packaged" software consists of all software written for multiple customers and for all types of computer platforms. It is software that is not written to individual customer specifications ("custom software") but for distribution through a variety of channels. All figures in this report are quoted in current U.S. dollars.

Table 5.3 Western European Software Segments (U.S.$ billion at 1996 exchange rates)

1996	Packaged Software	Professional Services	EDP Services	Network Services	Hardware Maintenance	Total
Value	39.54	32.81	9.78	4.01	18.92	105.06
Percentage	37.62	21.22	9.29	3.82	18.00	100.00

Source: Torrisi (1998).

Because the packaged mass-market software products offer complete customer solutions with few additional services required for installation, it has become the most rapidly growing segment.[15] In addition, drastic price declines and continuous technological advances have produced affordable desktop machines that are comparable to mainframe performance of just over a decade ago. It is no wonder then that U.S. companies have leveraged their technological advantages to enjoy dominant market shares across the globe in this market segment. In fact, U.S. firms continue to capture already existing dominant market shares throughout Asia.[16] Furthermore, with the recent emergence of widespread demand for e-commerce and Internet-related services, the continuing liberalization of telecommunications throughout Asia, and the development of intranets among corporations, the future belongs to those who can rapidly produce

Table 5.4 World Software Market Size

Region	Software Value	Market Percentage	Software/GDP Percentage	PCs/100 White Collar Workers
United States	145.47	42.4	1.87	104
Japan	52.76	15.37	1.08	24
Western Europe	105.09	30.62	1.14	88
Rest of the world	39.78	11.59	—	—
Total	343.10	100		

Sources: Torrisi (1998), p. 48.

standardized, packaged solutions for the global market. Applications to virtually any business of any size can now be made cost effective.

Although these trends reinforce U.S. competitive advantages, they may provide ample opportunities for European firms willing to adopt these new business models and work directly with their American compatriot firms.[17] With shortening product cycles in an industry that must be able to adapt quickly to a multitude of technological advances simultaneously, most software vendors have numerous alliances with original equipment manufacturers (OEMs) in a host of potentially complementary products. The well-known Microsoft-Intel partnership offers the paradigmatic example. While Microsoft focused exclusively on software, Intel concentrated on hardware. Each made numerous strategic alliances and acquisitions to build upon their strengths. They each had to be careful not to erode the value of their core competencies while commodifying complementary products.[18] Arguably, the need for collaboration is greater in the IT business than in any other.[19]

In this environment, it is not surprising that large European hardware producers like Siemens-Nixdorf have found ample opportunities to shift their expansion focus to their systems services business in cooperation with U.S. firms. Also, as software becomes increasingly complex to meet the sophisticated needs of diverse pools of users, Europeans firms find ample opportunities to partner with others in the provision of training, consulting, and services. Indeed, it is for this reason that the services business in Europe is growing at an even faster rate than the traded software business (see table 5.6). Not surprisingly then, many European software vendors choose to exit the software product market in favor of robust services growth. As a result, if consolidation was not feasible among larger European players, many of

Table 5.5 Mainframes as a Share of Total Hardware (%)

Year	1987	1995
U.S.	46.82	21.57
Japan	72.33	42.85
Germany	55.82	30.16
France	53.19	32.39
UK	53.41	27.74
Italy	66.11	44.64

Sources: OECD (1997), Torrisi (1998), p. 49.

the numerous, small, customized solutions' providers for specific industries were usually bought out by U.S. (or Japanese) firms intent on breaking into European markets.

The Asian Software Markets.   Despite the recent crises, the Asian IT market continues to offer enormous market potential. Given the large direct foreign investment commitments that American and Japanese firms and governments have made to the commercial infrastructure and labor base in the Asian region, the latecomer status of Europe is striking. With World Bank projections that the computer markets in Asian economies will outperform the rest of the world's regions for the next ten years, Asian IT markets do indeed entice Europe (see table 5.7). Furthermore, with favorable infrastructure improvements, low political risk, and high market receptivity and intensity, it is no wonder that Asia remains a key software market target (see table 5.8). In fact, East Asian software growth rate has matched that of the United States over the past decade, exceeding annual growth rates of 30 percent in some countries.

As table 5.6 indicates, not only is total computer spending growing at extremely high rates, the PC market is outperforming all other processor types combined (see table 5.9). This trend has been reinforced by the recent proliferation of the Internet and e-commerce as well (see table 5.10). As PC computer systems continue to get cheaper with the entry of several strong local competitors as well, it is not surprising that the U.S. mass-market software products continue to enjoy success in these economies. But it is also interesting to note that maturing business markets, primarily in Japan, have also led to the penetration of a client-server architecture based on medium-sized processors and workstations. Only

Table 5.6 IT Value Chain (percentage of total market revenues)

IT Market Segment	1992	1996
Vendor support and services	36	45
Distribution channels	11	7
Packaged software	16	24
Peripherals	12	8
Processors	14	12
Semiconductors	11	10

Sources: Siwek and Roth (1993), Torrisi (1998), and Hoch, et al. (2000)

the large mainframe-type processors have failed to post growth over the same fifteen-year period. However, the demand for high-powered processing involving massive parallel computing and mainframe systems with Internet and e-commerce introduction are likely to propel this industry as well. Such a situation bodes well for European enterprise solutions providers and professional service businesses alike, which have experience with large and medium-sized mainframe use, including workstations.

Nonmarket Factors

The Asian Nonmarket Environment.   As it becomes clear that IT use in organizations has moved from back office support to the center of what a firm needs to do to survive, it is not surprising that Asian governments believe that the software industry plays a critical role in their long-term economic prosperity. Consequently, they have made enormous, long-term commitments in infrastructure and education while taking active roles in the promotion of the IT industry. However, there are peculiar barriers to entry in this industry that are closely related to geographic proximity, access to key strategic partners' resources, peculiar consumer market bases, venture capital availability, and constantly expanding infrastructure requirements. Since there is no definitive work on the software industry, this has made it difficult for governments even to know, much less recreate, the conditions necessary for software success.[20]

But this has not stopped several Asian countries from taking dramatic steps to promote software production, including the provision of start-up

Table 5.7 Asia Computer Markets by Country

Country	Total Computer Spending, 1995 (U.S. $ millions)	Growth in Computer Spending 1985–1995	Total Computer Spending to GDP, 1995 (%)
Japan	96,590	9.5	1.89
South Korea	8,952	17.8	1.97
Taiwan	2,119	8.3	0.84
Hong Kong	1,887	17.1	1.31
Singapore	1,880	12.2	2.24
Malaysia	1,365	16.0	1.62
Indonesia	1,118	23.4	0.58
Philippines	573	27.0	0.79
Thailand	1,326	17.0	0.80
China	4,540	32.7	0.65
India	2,298	31.4	0.75

Source: IDC (1997).

venture capital and matching funds for industrial R&D, the creation of "Software Parks" that are designed to promote foreign technology transfer and training centers, the enhancement of university-business collaboration, the organization of standards consortiums, and stronger intellectual property laws and enforcement. There is a virtual consensus among Asian governments that technology transfer via Asian "Silicon Valleys" is critical to the creation of "local Microsofts." From Malaysia's Multimedia Super Corridor to Thailand's Software Park Project, all Asian governments have made concerted attempts to woo foreign investors to transfer their software development personnel and capabilities into their countries. In their zeal to promote this critical industry every Asian country has made attempts to: (1) streamline the business environment to make it free of government regulation, thereby encouraging direct foreign investment and/or strategic alliances; (2) offer favorable tax incentive programs and special amortization schedules for R&D investments; (3) strengthen copyright laws through coordination with international conventions, treaties, and organizations; and (4) create MOUs and policy commitments to step up the enforcement of intellectual property violations.

The two largest impediments to the development and expansion of the software industry in Asia are the lack of skilled labor (for both programming

Table 5.8 Ranking of Most Promising IT Markets (1997)

Countries	Growth Rate	Consumption Capacity	Commercial Infrastructure	Market Receptivity	Country Risk	Market Intensity	Overall Potential
Singapore	1	10	2	1	1	9	1
Hong Kong	9	8	5	2	4	1	2
S. Korea	3	6	13	6	2	14	3
China	5	1	19	20	9	23	5
Malaysia	2	13	11	3	6	20	8
India	10	7	1	21	15	17	10
Thailand	4	19	21	7	8	19	13
Philippines	12	13	12	8	17	11	15
Indonesia	6	5	22	16	12	22	16

Source: MSU–CIBER, "Market Potential Indicators for Emerging Markets (October 17, 1997).

Note: Dimension definitions: market intensity = GNP per capital, private consumption/GDP; consumption capacity = percent share of middle class in total consumption/income; commercial infrastructure = telephone lines, Internet hosts, TV sets, population/retail outlet, and paved roads per capita; market receptivity = per capita imports, trade/GDP; and country risk = Euromoney, March 1997, IMF. This survey included all member OECD countries outside of G-7 and EU countries.

Table 5.9   Computer Hardware Sales by Processor Type (Based on Value of Total Sales Revenue)

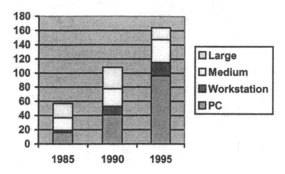

Source: McKinsey & Company (1996).

and technical support) and piracy. Because of the shortage of software engineers, the role of the domestic software industry in all Asian countries (except Japan), has been restricted to localization, customization, and installation. Even in the case of packaged software, localization costs are hardly minimal since they often involve the translation of highly complex Chinese-based characters. Manufacturing, sales, and marketing of software products require substantial training and expertise. Technical support requires familiarity with the functions and operations of the publishers' software products. Although this has led to the rapid creation of government-funded training facilities and computer science curriculums, Asian domestic markets are unlikely to develop to move beyond the role of bringing foreign products into their respective countries. Although such "localization centers" are reportedly multiplying in China, Singapore and Taiwan remain attractive locations for such regional hubs. However, the world labor shortage of computer programmers has led large software companies from the United States and Japan to outsource large scale, mundane programming tasks to the rapidly emerging cores of skilled software engineers from China and India. This is yet another area that Europe has been slow to capitalize upon.

However, for the foreseeable future, software "piracy" remains the major impediment to growth in the packaged software industry. Software piracy is the unauthorized production, copying, or distribution of copyrighted software products. Compared to the 26 percent piracy rates in the United States, which are among the lowest in the world, the Asia-Pacific

Table 5.10 Number of Internet Hosts

	United States	Japan	Korea	Singapore	Hong Kong	Taiwan
				Total Number of Internet Hosts		
1995	3,178,266	96,832	18,049	5,242	12,437	14,618
1996	6,053,402	269,327	29,306	22,769	17,693	25,273
1997	11,829,141	955,688	123,370	60,674	48,660	40,706
1998	23,178,266	1,687,534	186,414	67,060	82,773	N/A

	United States	Japan	Korea	Singapore	Hong Kong	Taiwan
				Number of Internet Hosts per 1,000 Population		
1995	12.2	0.8	0.4	1.8	2.1	0.7
1996	23.2	2.2	0.7	7.8	2.9	1.2
1997	45.8	7.7	2.8	21.7	8.4	1.9
1998	89.9	13.4	4.1	22.1	13.1	N/A

Source: Network Wizards (1995–1999). *Internet Domain Survey.*

region remains one of the world's worst violators (49 percent in 1998; see table 5.11). Although this represents steady improvement over time, some Asian countries ranked amongst the top in the world (Vietnam 97 percent, China 95 percent) while others leave significant concern for improvement (Indonesia 92 percent, Thailand 82 percent, and Malaysia 73 percent). Of the global estimated total of piracy losses of $11 billion, the Asia-Pacific accounts for the second highest regional total at $3 billion for 1998.[21] With a single country total of close to $1.2 billion, China is the single largest violator, followed by Japan ($597 million) and South Korea ($198 million).

Studies have shown that the legal protection of software, combined with strong enforcement and increased public awareness, results in a dramatic reduction in software piracy.[22] Preventing illegal copying not only accelerates job growth in the host country but also increases the tax revenues generated from the packaged software industry. But this is only one more peripheral incentive governments get from intellectual property enforcement. Through becoming part of the Berne Convention and the World Intellectual Property Organization's (WIPO) copyright treaties, Asian countries have sought to promote their own domestic software industrial development while becoming part of the emerging global information infrastructure. Because of the public good qualities associated with software package development, there is evidence to suggest that programmers require the ability to protect their innovations, the right to protect how their works are sold and distributed as well as the support of public funds to carry out the expensive and risky R&D associated with this sector.[23] Without the ability to exploit the growth opportunities associated with the generalization of these technologies, the market may not provide the necessary incentives to promote innovation. In fact, there is general consensus among many analysts that the U.S. software industry may not have achieved world leadership in computer software and services without the comparatively early development of a domestic copyright system for software technology.[24]

With significantly declining piracy rates throughout Asia, it is clear that governments in the region have taken these lessons to heart. In addition to providing the market incentives necessary for software development, the provision of formal education and the enforcement of intellectual property rights continue to make Asian economies an attractive target for software firms. As a sector where rules of origin and local content requirements are generally relaxed, many exporting opportunities will continue to arise on a consistent basis. Finally, with the recent progress in the

Table 5.11 World Software Piracy Rates and Losses, 1994–1998

	Piracy Rates (%)					Retail Revenues ($ millions)		
	1994	1995	1996	1997	1998	1994	1996	1998
Total Asia-Pacific*	68	64	55	52	49	3,144.5	3,739.3	2,954.8
China	97	96	96	96	95	364.0	703.8	1,193.4
Hong Kong	62	62	64	67	59	64.5	129.1	88.6
India	79	78	79	69	65	103.1	255.3	197.3
Indonesia	97	98	97	93	92	104.5	197.3	58.8
Japan	66	55	41	32	31	1,399.8	1,190.3	596.9
Korea	75	76	70	67	64	510.6	515.5	197.5
Malaysia	82	77	80	70	73	66.7	121.5	79.3
Philippines	94	91	92	83	77	40.6	70.7	31.1
Singapore	61	53	59	56	52	37.3	56.6	58.3
Taiwan	72	70	66	63	59	112.0	117.0	141.3
Thailand	87	82	80	84	82	67.8	137.1	48.6
Vietnam	100	99	99	98	97	3.9	15.2	10.3
Total W. Europe	52	49	43	39	36	2,783.0	2,574.9	2,760.3
Total N. America	32	27	28	28	26	3,931.1	2,718.3	3,195.8
United States	31	26	27	27	25	3,589.5	2,360.9	2,875.2
Total World	49	46	43	40	38	12,346.5	11,306.3	10,976.5

Source: International Planning and Research Corporation (2000).
*The Asia-Pacific also includes Australia, New Zealand, and Pakistan.

WTO's International Technology Agreements (ITA), tariffs on thousands of IT products, which are already comparatively low, will be completely phased out.

European Policy Responses. Even earlier than most of their Asian counterparts, European policymakers and practitioners have long understood the strategic importance of software to such industries as banking, airlines, automobiles, insurance, and publishing as well as virtually all consumer products.[25] A recent EU survey estimated that 70 percent of all software developed in Europe will be for the non-IT sectors of the economy.[26] As mentioned before, the peculiar characteristics of this industry have left European firms in catch-up mode for a moving target, the stand-alone packaged product.[27] In particular, the emergence of the "network" or "virtual" firm to adapt to the growing complexity and uncertainty of business activities represents a radical departure from the vertically integrated model.[28] This model continues to be widely seen in the software industry, which is dominated by U.S. independent vendors in alliance structures. In addition, there is a strong path dependent effect related to the lock-in effect of users trained in particular software interfaces and developers trained in particular languages and codes. For example, the amount of code in most consumer products and systems is doubling every two to three years. As a result, software developers struggle with coping with the pressures of systems that are not only a couple of orders of magnitude larger and more complex than those developed a few years ago but also have to meet ever-increasing demands for quality and performance.[29] Furthermore, studies have shown that the high R&D costs associated with software development indicate both a high failure rate and spiraling budgets.[30]

In view of these recurrent problems in software development and the perceived central importance of the software industry, there has been both a European-wide technical and managerial response. In the words of the current European software policy:

> As the information age develops, software will become even more pervasive and transparent . . . the ability to produce software efficiently, effectively and with consistently high quality will become increasingly important for all industries across Europe if they are to maintain and enhance their competitiveness.[31]

These statements linking software with overall European competitiveness are taken from the first page of the European Systems and Software Initiative

(ESSI), established in 1993 by fourteen leading European companies with support from the European Commission and the government of the Basque region of Spain. Focusing mainly on the organizational and management challenges of producing software, ESSI seeks to create a cooperative consortium of companies to avoid duplication of effort, unnecessary competition related to incompatible standards, and the promotion of best management practices. This initiative was designed in coordination with other EU-level efforts of a more technical nature such as Specification and Programming Environment for Communication Software (SPECS), a representative Research and Development in Advanced Communications in Europe (RACE) project. SPECS is a project designed to develop common methods and techniques for European-wide coordination of the development of the complex software needed to create the pan-European integrated broadband communications (IBC) system. The approach is to use formal methods and maximum automation to support the multiple specification languages currently in use in various European countries across several industries.

To put SPECS in perspective, the IBC is the modern telecommunication infrastructure to usher in the twenty-first century system of terminals, cables, switches, computers, and satellites that handle telephone, television, data transmission, and services in an integrated way. RACE is a program funded 50.1 percent by the EU to design the architecture and provide the technology necessary to build it. Finally, SPECS is designed to meet the challenge of providing the extensive, standardized and complex software necessary to make IBC operational. The central idea is to create SPECS tools such as formal methods to enhance productive efficiency in an open environment. The development of common tools and methods in design, implementation, testing, execution, maintenance, and adaptation promises has created a cooperative spirit among the various telecommunication, data processing, and software developer participants that work at making the IBC a reality. Furthermore, several long-term efforts at providing matching EU venture capital funds, simplified and harmonized pan-European software procurement practices and the development of multilingual modules have all helped European software firms collectively coordinate their catch up efforts through sharing risk and R&D costs.

Yet despite these significant governmental efforts to protect and promote European software firms for close to a decade, the results remain unclear.[32] Not only do government funds (from the EU and national governments alike) tend to go to established companies working with universities, other sources of venture capital funds are still hard to come by for start-ups seeking to create stand-alone packaged products.[33] Preliminary evidence suggests that the lion's share of money from national software promotion programs go to large established companies in the form of contracts, grants, and capital.[34] The information services and software divisions of large telecommunications giants have been the main beneficiaries of government procurement, standards bodies' coordination, and other institutionalized integration efforts. Because they are primarily domestic producers for large local market niches, such as banking, the supranational EU institutions, national government agencies, and local municipalities, they generally lack either an export-orientation or direct foreign investment commitments. But despite the European-wide consolidation over the past two decades that resulted in fewer, larger software firms, more innovative foreign, packaged, traded software products continue to cut into these higher-end domestic market shares as well. As a result, one study even found that nearly 20 percent of Europe's 50 top software companies had moved their headquarters to America.[35] In part, this is due to the fact that Western European labor productivity in the computer and communications technology sector lags 41–55 percent behind their American and Japanese counterparts.[36] But it is primarily due to the need to work closer with complementary technology partners in an industry with shortening product cycles.[37] It has become painfully clear that even for niche markets, globally packaged solutions compatible with multiple environments are the wave of the future. What was once a comparative advantage—a large, diverse, domestic market serviced by specialized customized solutions—appears to remain a latecomer disadvantage to entering these evolving software product segments.

III. Strategic and Tactical Analysis: European Software Firm Experiences in Asia

Market Strategy and Tactics

Software product development is a creative, craft-like, labor-intensive activity with negligible manufacturing costs. Fixed development costs are high and performance uncertainty remains constant throughout the testing phases. But unlike customized software projects done on a case-by-case basis, packaged software products can take advantage of economies of scale with the emergence of a market standard platform for PCs and distributed client-server network architectures. Because replication costs are virtually zero, a software product is costly to produce but cheap to reproduce. As discussed earlier, this has brought about a rapid proliferation of firms specializing in packaged software applications. Because of the declining costs of both the PC platform, the dramatic increase in processing power, and the rapidly expanding base of consumer and business applications alike, these fast growing packaged software programs, like Microsoft, Lotus, and Novell took dominant market shares in Asia so quickly that the initially high estimates of piracy did not deter aggressive entry strategies.[38]

As an industry closely dependent upon computer hardware (and a host of closely related technologies), this U.S. competitive advantage was further reinforced by the network externalities, which arise from the proliferation of complementary software and hardware products. Not only does such innovation in a host of related technologies expand the size of the market, it creates further opportunities for increasing returns. These market forces have led to a high market concentration that is similar to the hardware business, where most PC producers set up contractual linkages with co-specialized suppliers of software applications and tools as an alternative to in-house development.[39] The well-known example of operating system concentration with IBM at the top end, Microsoft Windows on the desktop, Sun's Solaris for client-servers, and Windows NT in the middle is an accurate depiction of the relatively few global producers we see in the major software product segments. These network externalities worked against the major European software producers, which were originally hardware manufacturers offering highly complex customized software and services, leaving them at a comparative disadvantage in this expanding global segment.

This is primarily due to the fact that the economics behind customized software and services is strikingly different from the mass-market packaged market (see table 5.12). The production of customized software goods involve high marginal costs that depend on end-user requirements and specifications. This is related to balancing the mix between reusing the clients' existing designs, source codes and methodologies with new specifications, objectives, and equipment.[40] The rapid evolution of computer systems (and related technologies) usually combines with new user requirements to allow highly inefficient reuse rates. Also, computer services require a high degree of close and repeated interactions between production and consumption, making it difficult to separate products from processes.[41] In addition, user-supplier interaction increases fixed costs, which are variable to

Table 5.12 Software Dynamics in the Three Major Market Segments

	Customized Services	Enterprise Solutions	Packaged Products
Marginal costs	Almost constant	Variable by client, project	Almost zero
Market structure	Highly fragmented	Concentration tendency	High concentration
Appearance	Regional	Global tendency	Global
User relationship	One to one	One to one or few	One to many or few
Key indicator	Capacity utilization	Installed base, efficiency	Market share

Source: Adapted from Global McKinsey software survey, Hoch, et al. (2000). p. 46.

a specific client's needs, specifications and requirements. Although increasing returns to scale are important in this business as well, it is dependent upon the size of the contract rather than the volume of production. This helps to explain the bifurcation in the customized software market where a handful of large multinational services suppliers (such as Cap Gemini and EDS) that offer total solutions for larger clients coexist with numerous small firms that focus on specific local market niches.

Without extensive licensing arrangements with key OEMs, European firms have found it extremely difficult to enter the mass-market for packaged software products. However, because of the penetration of the PC in their own markets, demand for new business applications came to the forefront of software activities, leaving wide open the market for enterprise resource planning software. As mergers and acquisitions flourished and international supply chains expanded, standardized decision-support and online database management systems were in high demand. Utilizing the low cost PC and workstations, this demand was increasingly met by the widespread diffusion of distributed computer architectures (e.g., client-server networks) that became responsible for the diffusion of new packaged solutions as well as the distribution of information processing within large organizations.[42] SAP was the first company to bring a high-quality, reliable product to large companies in this rapidly growing software product niche.

Several European software firms were finding ways to leverage their customized solutions' and services' strengths to create complex, standardized products for targeted groups of users, including those with expanding operations in Asia. In particular, new emerging markets in Asia were especially attractive not only due to their impressive track record of growth but also since they did not have to resolve fundamental system integration issues as entirely new systems could be installed from scratch. It is in this way that European producers, such as SAP and BAAN, were able to enter the global software product market—in the product-services hybrid, the enterprise solutions segment. Unlike mass-market packaged goods, enterprise solutions software require substantial time, effort, and resources to get up and running. Because they involve connecting back office functions like sales, payrolls, procurement, distribution, finance, and inventory, implementation is a complex process that links together multiple, heterogeneous platforms (often in different geographic regions). As full installation and training ranges between six months to two years, the enterprise solution product is designed for the thousands of large companies, not the multi-million masses. For example, while Microsoft sold sixty million copies of

Windows 95 in 1997 alone, SAP, which continues to be the world's leading enterprise resource planning (ERP) software provider through 2000,
installed its R/3 product at approximately 23,700 sites over a period of five
years.[43] ERP implementation is also prohibitively expensive. In order to
make systems compatible with multiple standards, a degree of customization is always required. One study showed that companies that install ERP
software end up spending an average of 70 percent of the total cost on professional services to implement the product.[44] While competitor Oracle
offers these services in-house, SAP and BAAN rely on local partners for
implementation. The $60 million Swiss enterprise solutions firm, Simul-
tan, is more typically the size of an ERP full service provider, which develops industry-specific solutions for about fifty customers a year.

However, enterprise solutions are not the only option for European
firms seeking to get in on the growing software market. In fact, because
Western European IT markets are a patchwork of languages, legal and
accounting systems, organizational cultures and business processes, it is
not surprising that the large Western European IT market has remained
primarily a customized, professional services business. With a host of
prior expensive technology to make use of an installed base of enterprise
users, the dramatic rise of PC use and decline in the mainframe have not
offset the need for systems integration nor customized solutions. In fact,
the strong dependence on large systems and customized software led
many European firms to specialize in helping large organizations cope
with the increasing complexity associated with maintaining and merging
large information systems in consolidating clients and their overseas operations. Particularly, demand for facility management, systems consulting, and system integration became so great that five of the top eight
European firms selling more than $100 million in software and services
in 1994 were all selling specialized software to large companies. Other
large players such as Cap Gemini and Debis chose to leverage their previous experience and capabilities by specializing in custom programming, professional services, and consulting. However, because this is
primarily a regional business where in-depth knowledge of clients' business and prior technological expertise are the key assets, their products
are not readily packaged for widespread consumption.

In short, to do well in Asia, European software producers must go beyond the continuing exploitation of their specialized customized software
capabilities and expertise associated with adapting older programming languages to new applications. Rather, success in Asia involves taking advantage of the huge opportunities for economies of scale in the production of

packaged application solutions based on both the PC-compatible platform as well as the distributed computer architectures, such as client-server networks. The successful track record of European enterprise solution products illustrates the importance of realizing that serving a global market is the key to survival in this business.

But given the prohibitively high development costs, the traded software market requires ready access to capital and deep pockets to outlast intense competition for market share. In fact, as early as 1985, the OECD reported that software costs accounted for over 80 percent of the total (including hardware) costs related to new information technology applications.[45] A recent survey conducted in the Netherlands found that 23 percent of total business R&D expenditure was accounted for by software in 1994.[46] Thus far, few of the European IT firms have been willing to make the long-term investments necessary to be successful in this business. For example, of the world's twenty largest IT firms ranked by R&D intensity (of which five were software producers) in 2000, only one, SAP, was European.[47]

However, because many European firms do have an important closely related software service niche from which to draw revenues, they are in a solid position to take advantage of new technologies when they arise. Netscape, CSC, and IBM are all successful examples of hybrids that produce standardized software, hardware, and information processing services. As capabilities develop in Asian infrastructures, labor pools, and other foreign partners, creative partnership opportunities have been crucial to lowering time to market for new application products. In an industry with rapid product cycles and high bankruptcy rates, many opportunities to enter the market are likely to arise on a continual basis, especially with the recent introduction of component ware, new Internet applications, and wireless technologies.[48] The new challenge of this network area will be to build marketing alliances, create software on new platforms compatible with old technologies, and continually innovate by launching a steady stream of new products and services. This means that cooperating with local partners and managing a host of informal, equal partnerships abroad. Taking these lessons to heart, SAP realized early on that it would need to set up training institutes in multiple locations around the globe to ensure that implementation would not become the bottleneck to entering Asian markets. Accordingly, SAP set up its innovative Partner Academies throughout the region, providing technical training to its "clients and partners" in dozens of weeks long courses. Such a facility was the first of its kind, dedicating enormous resources to its local partners in the process. SAP realized early on that the software implementation services business was about trust. Since a SAP ERP module is expensive and sold before any result is delivered, they needed to find ways to share implementation risk as well as demonstrate a commitment to finish the project completely and in a timely manner. By partnering with local software localization centers (rather than provide these services in-house), the sufficient market incentives between alliances of equal partners meant that both would have an economic interest to see the project to its fruitful end. Beginning in 1997, SAP further consolidated and expanded their training facilities in Asia through the creation of full service IT technology and management programs in a more comprehensive curriculum (called Sapient Colleges) in Hong Kong, Indonesia, Malaysia, Japan, South Korea, Singapore, Philippines, Taiwan, China, India, and Thailand.

Thus, given the previous competencies of European software firms, only the enterprise solutions providers with global products have become successful global players in the Asian market. But as network software applications have grown with a large, installed base that exists on a global level, demand for integrated application solutions, which include accounting, human resource management, manufacturing, and the like, is likely to remain robust. Although many enterprise solutions clients are primarily for blue chip companies in the United States and Japan, an increasing number of Asian subsidiaries and local producers have also found ways to take advantage of this software product. While SAP became the world leader in applications software for business computer networks, Germany's Software AG built a $400 million turnover in the high-end database tools business. Other firms like Micro Focus and Synon found niche markets in the same manner, namely by leveraging their expertise in producing developmental software tools that made it easier for firms to create software making tools for various platforms.

Although they began consolidating market share in their own national markets first, European companies like SAP were successful because they were early in bringing their product to the global market and aggressively pursued market share through extensive local partner alliances. Although initially finding it difficult to break into the Japanese market with strong corporate loyalty to their long-term computing partner, economic recession and the recent Asian crises have further helped SAP make significant inroads in the ERP market. After consolidating the high end market throughout the U.S. and European IT markets, they then began attacking the small- to medium-size enterprise that is more relevant to the less developed Asian markets. In a hotly contested market that competes with the likes of Oracle and Peoplesoft, SAP has consistently demonstrated the ability to tailor their standardized solutions for diverse industrial clients that requires intimate understanding of specific business operations, regulations, and informal business practices.

Recent trends suggest that demand for the software services business will continue to remain steady as the localization requirements for implementing systems in heterogeneous environments remains increasingly intense.[49] However, even short-term demands such as the Y2K bug and Euro currency consolidation have pushed software companies toward creating integrated packaged solutions that can serve a global market. The "innovate or die" slogan means that the lion's share of the profit will flow to anyone who can offer the best standardized software that can be used by the largest numbers of organizations.[50] As relationship-based markets where reputation and brand are paramount, software producers who are durable and willing to invest for the long-term will do the best in emerging markets such as Asia over the long haul. Providing stand-alone products that provide clients with complete solutions and around-the-clock customer service, which are compatible also with complementary products and scalable over time, are the key competitive factors in penetrating Asian markets.

Nonmarket Strategy and Tactics

Of course, all of this is not lost on European governments seeking to assist their own domestic software industries. European governments' ambitious plans to create a cutting-edge information infrastructure based on developing innovative European IT and software companies indicate a long-term commitment to competitiveness and more concretely, their own software products and modules.[51] This has opened up widespread opportunities for national telecommunications companies, electronic conglomerates, custom software service providers, ISVs, and new European entrants alike to enter the software business in a captive market. Similar to the early days of the development in the U.S. software industry, governments are likely to be guaranteed markets for their expensive, specialized products. But as discussed earlier, these customized projects may not be readily suitable for widespread consumption. Developing complex systems for large-scale organizations like governments and telecommunications puts a premium on reliability and specialized user requirements that may not be readily transferable to the other organizations. However, like the Japanese, the Europeans believe that software can be divided amongst a larger division of labor.

In line with this strategy, the government has put a premium on supporting projects that focus on the development of tools that make it easier to make software for clients; and on modules, which are pieces of programs that will be interchangeable to diverse industrial settings and functional purposes. To date, the classic pattern for European software producers has been to utilize the nonmarket incentives available to consolidate their market share in their national markets first. Then, taking advantage of the available EU funding, entering other European markets through offering menus, manuals, and help services in over a dozen languages by utilizing local distributors and technical conventions to account for distinct national and organizational preferences. Although much of the work in creating an information infrastructure is highly specialized, it is also potentially transferable to a larger, emerging global market. The key competitive dynamic for an ISV is to be able to standardize either a developmental tool or a specialized application product to a clearly defined set of users. Especially in Asia, governments remain some of the biggest clients in the world in search of such scalable solutions. Given the present market structure and capabilities of European firms, Richard D'Aveni's firm strategy prescriptions to find ways to disrupt the market through creating new ways to satisfy customers makes sense.[52] As a follower in a rapidly evolving market, finding ways to leverage one's existing customer base through creating globally packaged solutions designed for speed and surprise is exactly how the leading European software companies, SAP and Baan, came to dominate a growing software niche.

But European software firms also face a favorable nonmarket environment in Asian countries as well. As a large body of literature suggests, doing business in this region is heavily influenced by both the state's industrial policy and its regulatory environment. While industrial policy can change the incentives of private companies and individuals, alter the competitive environment of firms, and assist in the creation of national capabilities, the regulatory environment may present a host of nonmarket risks to foreign firms. The latter usually involve several nontariff barriers, including changes in the timing and availability of export licenses, tariffs and other trade barriers, difficulties in staffing and managing foreign operations, potentially adverse tax consequences, difficulties in obtaining governmental approvals for products, distributor mismanagement, and changes in regulatory requirements. However, in the case of packaged software, Asian governments have made comprehensive efforts to make the regulatory environment highly efficient and transparent to encourage foreign software company direct foreign investment, licensing, and localization efforts.

Not only do they usually offer extensive tax breaks, relocation incentives, and precompetitive research grants, but they will often provide assistance in locating qualified personnel as well as participation in R&D consortia.

As governments continue to devote large resources to provide the specialized facilities, electricity, and other infrastructure requirements, industrial clusters such as the thousands of computer and electronics companies in Taiwan's Taipei-Hinchu area and Singapore's strong supplier base of skilled engineers in the disk drive industry have continued to make these attractive locations for software companies to locate. For example, in Taiwan, foreign firms that participate in precompetitive research in majority-owned Taiwanese joint-ventures will also become subject to both lenient bankruptcy laws and government subsidies. Furthermore, the Taiwanese and Singaporean governments have been known to work directly with multinational corporations (MNCs) to set up joint research centers with IBM, Apple, Siemens, and SAP.[53] Seeking to be hubs for the IT industry, the central core of Singapore's strategy has been to get the leading MNCs to locate in Singapore, where government officials work as account executives to keep them happy. Taking a different tack, Taiwan sought to promote export through the provision of extensive market research. Taiwan's Institute for Information Industries has become an important source for providing accurate and timely market intelligence for MNCs.

In part due to their success, Japan and South Korea have also recently begun to open up their coordinated high-technology R&D consortia involving government labs and private companies to foreign software companies. But informal distribution channel barriers, expensive labor, and high land prices have kept many companies from locating in Japan beyond the creation of localization centers; and in South Korea, informal barriers remain strong in several strategic IT industries, consistently subject to challenges at the World Trade Organization (WTO).

In short, although various national Asian software promotion schemes exist for established producers, there are a host of nonmarket factors that must be weighed against the market environment. A user-friendly government can go a long way in working through the maze of business permits, regulatory approvals and overall flexibility needed in the course of meeting market demands in a highly competitive global industry. It is no coincidence that Taiwan and Singapore have become software localization center hubs, while such investment has lagged in Japan and South Korea, the two largest markets. Because the latter countries focused on promoting the computer hardware promotion over the application of IT to business, government, and education, they have lost many opportunities for developing software and services industries. Today, all Asian countries recognize the importance of the software industry and have pushed for more open policies to MNCs in this regard. With such a favorable nonmarket environment that provides capital, training, marketing intelligence, opportunities for procurement, and streamlined regulatory environment, it is no wonder that European enterprise solutions have found some success in Asian markets.

IV. Corporate Strategies: Case Studies

The Overall Context

Given its high profile nature, the lack of published scholarly research that provides empirical studies of European software firms might seem surprising.[54] Yet given the inherent difficulties of accurately analyzing and measuring the processes of an industry currently undergoing radical change, such an omission is understandable. This section seeks to help fill the empirical gap in the literature by elucidating the experiences of a few European firms in the enterprise solutions business. The analysis of the market and nonmarket environments of the previous sections will provide the tools for this approach.

As a latecomer to the industry, the prescription that European software firms should focus on disrupting the status quo through a series of temporary advantages is apropos. In addition, D'Aveni's prescriptions for firm market strategies can also be interpreted through a nonmarket lens.[55] Envisioning a disruption can involve creating strategically placed niche products as much as lobbying the government to pursue a competitor for anti-trust abuses. As David Baron has clearly articulated, understanding the nonmarket environment is equally important to formulating firm strategy as market dynamics.[56] As we will see, the few European firms that have been successful are masters at implementing an integrated strategy for multiple platforms.

As previously discussed, the availability of software increasingly drives the production of hardware products in computer and microelectronics. Because growth continues to flow from widespread adoption in diverse industrial application settings, there is ample opportunity to export a standardized solution to markets with a similar base of installed platforms and users. In the case of enterprise solutions, there is also significant opportunity for those who can provide the complete solution to a large firm. But they must be able to keep up with the technological innovations that affect the dominant platform designs. In part, this means that European producers must stay in tune with a host of complementary technological changes occurring in the most sophisticated and largest software market, the United States. In fact, providing those U.S. firms that have an established presence in Asian markets with software solutions is one of their most important entrees into Asia.

Although national customized market niches, which provide important sources of revenue, will continue to grow incrementally, it is not clear that they will survive the next wave of technological products. Neither is their projected growth anywhere near the levels expected in the packaged market sector. A survey of European firms found that over half of the respondents expected their custom software markets to decline in the near future.[57] With the advent of new computer languages and object-oriented programming, software firms whose core competencies lie in the packaged market may eclipse them entirely. In fact, internationalized packaged producers have demonstrated the ability to exploit both economies of scale and scope in the maturing Asian markets. As existing East Asian international production networks (IPN) continue to generate world class original equipment manufacturer (OEM) and design (ODM) facilities through multinationals' foreign direct investment (FDI) in cooperation with host governments,[58] infrastructures continue to improve and labor bottlenecks become less of a concern.

While most have long since exited the market to focus on services and consulting, a few European firms have adapted to the new hypercompetitive dynamics of the international software market. Of course, this exit by both major hardware players and independent software vendors makes sense from a purely market-based strategy. As mentioned before, services are a booming industry with expected growth levels that exceed even packaged software. Particular factors, such as the Y2K bug and the Euro, created short-term incentives for developing software tools and standardizing automated accounting processes. But the long-term competitive dynamics reflect a high turnover rate associated with rapid technological advances and a constantly evolving installed base. Under these circumstances, most companies do not want to be dependent upon a customized solution that will become quickly obsolete from a company that may not exist in the next technological product cycle.

As opposed to services, software restructuring is a highly costly process. It is not easy to package one's own custom software to multi-client markets. New developmental tools such as object-oriented programming and fourth generation languages (4GL) may offer distinct advantages since they expand reusability of software modules for different uses, thereby lowering entry costs. Large manufacturers like Olivetti, SGS-Thomson, and Groupe Bull have all undertaken radical restructuring and reorganization to focus on services, components, and telecommunications to maintain their IT presence. Government-induced cooperation has facilitated early harmonization of standards to avoid unproductive duplication and cutthroat competition. A steady stream of government procurement contracts have also helped create positive incentives to cooperate for bigger players apt to let the market decide. In addition, the EU has also assisted in technology transfer efforts by sending out bids to leading software firms for government jobs and linking them with local partners. This allows major software producers like Cap Gemini, the Sema Group, and Logica, who have all been concentrating more and more on servicing the faster growing packages software market segment, to get up to speed with the latest technological innovations. It is much easier to teach people how to use and adapt packaged suites to fit their needs than to undertake the huge R&D gamble that software production has become.

In addition, it is also simpler to leverage the local market knowledge advantages that have made the software niche business so lucrative in the not so recent European past. The localization, implementation, and adaptation associated with modifying existing standardized solutions involve less risk, less investment, and higher profit levels. Even with both national and EU policies to help spread the risk of developing software (from both a technical compatibility standpoint and capital provision requirements), the recent Internet explosion and U.S. innovative competitiveness have only underscored the uncertain nature of this endeavor.

But more specific to the theme of this volume, what types of obstacles have European software firms faced in breaking into new markets in Asia? We have already mentioned the increasing returns' factors associated with having dominant de facto standards and a large installed base further reinforced by technological developments such as networks and the Internet. The primary barriers to entry in the systems, services, and packaged software markets are: (1) the lack of in-depth knowledge of users' needs which are normally developed through long-term relationships with customers; (2) extensive marketing and distribution networks; (3) long-term access to capital; and (4) reputation, which is particularly crucial to potential clients. Recent surveys conducted over the past decade have indicated the importance of reputation as the consistently most critical factor.[59] This advantage alone has enabled the dominant U.S. players to overcome their deficiencies in other areas, particularly the establishment of long-term relationships.

Like highly internationalized firms (with the exception of Microsoft), most European firms are not highly diversified, but are mostly consolidated in specific market niches. As such, the historical development of specific patterns of specialization of European firms appears to be their primary competitive advantage to consolidate and create a wedge into their own U.S.-dominated packages and systems markets. This led to the tendency of locating internal R&D activities in countries that have a comparative advantage in these activities.[60] Not surprisingly then, there is a tendency for European firms to "turn American," not only for market access objectives but for venture capital as well as end user and producer interaction concerns. In addition, learning from their American and Japanese counterparts, European firms have also been able to tap into the skilled labor pool of software engineers, value added resellers, and local consultants.

However, a recent survey of European firms found that most continue to attach little attention to R&D, citing that the vast majority of their business is related to incrementally modified products rather than entirely new ones.[61] They are more concerned with—in order of importance—identifying clients, monitoring competitors, and the hiring and training of personnel. In fact, statistics indicate that from 1989 to 1992, only one-fourth of all firms derive the majority of their sales from totally new products.[62] With the exception of SAP and possibly BAAN, this is a trend that has continued.[63] In this sense, firm strategies and the organization of innovative activities generally remain consistent with the specialization of the European software industry in custom services and software. There are indications that this will continue to be a viable strategy for persistent niche markets, but also evidence that such companies are likely to disappear either through buyouts or mergers.

For the European software industry, this is a disturbing trend given the increasing importance of software to continuing economic growth and development. Although its potential ubiquity in virtually every business sector integrated into a global information highway remains a vision of the future, it is one that practitioners and policymakers all over the world dare not ignore. To some extent, competitiveness has become synonymous with software and IT. Yet it is still not clear that national and regional programs, policies, and initiatives will be sufficient to propel an independent software industry in the long run. The next section will examine three case studies of successfully integrated global strategies of European firms. We shall see that they all share four characteristics: (1) a global strategic vision; (2) skill and creative leveraging of regional advantages; (3) building upon their path-dependent organizational resource capabilities to leverage specific market niches into an integrated strategy; and (4) sensitivity to the non-market institutional environment of Asia.

Market Entry Concerns

Two central questions must be addressed before determining the entry mode into the higher risk area of the Asia-Pacific. First, what level of resource commitment are European firms willing to make; and second, what level of control over their operations do they desire. When risk is high, there is a reluctance to invest resources and rely on local contractees.[64] Conversely, when risk is low, there is a tendency to desire more control and run a wholly-owned subsidiary. However, in highly competitive industries, where many competitors exist and entry and exit are easy, managers perceive control to be a higher risk; both because there is much more competition and because the actions of the large number of competitors tends to be unpredictable. In addition, in highly competitive industries, as the pool of trained and knowledgeable managers in the country increases, the chances that a firm can obtain local contractees with sufficient knowledge and skills to reduce the need for control increases.[65] Also, the transaction cost of using non-integrated entry models decline over time.[66] Therefore, in highly competitive industries firms will likely use low control modes to expand internationally.[67] The case studies that follow illustrate the validity of these conclusions.

Case Study 1:
SAP—Continuing Global Evolutionary Success

SAP has become the world leader in ERP software, offering a full line of applications software for business computer networks, one of the hottest growth areas in office computing in the late 1990s. SAP is now a thriving $5.3 billion company that continues to make inroads into the enormously profitable $23 billion ERP subindustry in 1998, valued at $42 billion for 1999.[68] Through the first three quarters of 2000, SAP ERP solutions held market share three times as large as its nearest competitor. In addition to its domination of software for client-server ERP programs, SAP controls over one-third of the market for software used to integrate and process information in corporate product distribution, finance, human resources and manufacturing. It is the only European company to have a majority market share in any software subcategory and continues to roll out new packaged suite offerings tailored for specific types of businesses. By the end of 2000, SAP remained the third largest independent software supplier company in the world and the only European producer with a wide scale presence in Asia that continues to grow and expand throughout the region. With a strong commitment to training and software localization through the development of local value added resellers to implement their complex systems, SAP continues to grab market share despite the V-curve effect of the Asian Crises. Finally, with its current expansion into connecting front office e-commerce functions with the back office ERP software modules, Asia remains the core component of their strategy, growing at 40 percent through the first quarter of 2000, compared to 15 percent growth in Europe and the Middle East and a declining 3 percent rate in the United States.[69] Using Japan's SAP Labs and Singapore as the key Southeast Asian headquarters, SAP has rolled out an ambitious regional expansion program that has focused on both MNC and publicly-owned enterprises in large-scale industries, such as oil, chemicals, telecommunications, government, banking, and utilities.

The original impetus for SAP's creation occurred in 1972, when five engineers working for IBM Germany were not given permission to work on developing an order-entry program they wrote for a customer. It was a vision of global market integration that has propelled SAP into the top ranks of the independent software vendors of the world. But why has SAP been successful while all other European software players have fallen by the wayside? Like their European counterparts, SAP began as a company offering a customized solution to a specific application problem—reducing time to market through an integrated order-entry program that could work across multiple platforms. In consolidating the domestic European markets, they also followed a familiar IT trend, focusing first on reengineering and saving money by streamlining operations.

But unlike other European software developers like Cap Gemini, they chose not to specialize in the area where their competitive advantage of local market knowledge could best be exploited—the lucrative services and consulting businesses. Instead, they chose to attack the innovation bottleneck itself through creating packaged software products designed for a global marketplace. This meant having a broad "market-disrupting vision" that foresaw the reengineering mania that took over corporate America in the early 1990s.[70] One of SAP's newest line of flagship products, the R/S (and later R/3), is a software package that integrates a company's accounting, payroll, supply-chain management, marketing, and other activities on networks of personal computers. By providing detailed and accurate information about all of the firm's operations throughout the world in a central location, it helps companies of all sizes to become more efficient, flexible, and responsive.

A key part of SAP's strategy has been to focus on disrupting the status quo through a series of temporary advantages to provide itself time to build the capabilities needed to quickly capture opportunities and surprise competitors.[71] Furthermore, they developed tactics to shift the rules of the game in their favor by leveraging them through strategic signaling as well as simultaneous and sequential strategic thrusts. All of these elements can be found in SAP Co-CEO Hasso Plattner's keynote speech at the high profile Enterprise 1998 confab for top ERP companies. In Plattner's words:

> We dominate the most important category of enterprise software. We intend to control all the enterprise software our customers use. We will select a handful of partners to work with. If our partners cross us, we will crush them into dust.[72]

SAP saw the need for companies with global operations to go from procurement to customer in the shortest possible time through the fewest organizational points.

In short, SAP first established a presence in the global market by producing a client-server product that solved a central problem that virtually all major multinational enterprises had—transferring information and data from multiple environments in numerous countries to ensure operation and financial efficiency in real-time. Although the majority of their revenue originally came from its home base, Germany, it has since diversified its revenue sources. In 1999, the United States accounted for over half of its revenue (more than double that of Germany) and one-fifth of its shareholders, while Europe accounted for 27 percent. Even though Asia/Australia comprises only 9 percent of the current market, it grew at a faster rate (52 percent) than any other single region in 1999 (e.g., Europe 46 percent, America's 12 percent). As stated earlier, this trend has continued through 2000. A central key to its market strategy is to partner with competent local partners to serve more customers quickly. The cooperation of Singaporean and Taiwanese governments has been crucial to not only procurement contracts but with private partnering opportunities and the provision of favorable arrangements for their local training centers. Rather than participate in the trend of buying out smaller rivals, it has found ways to collaborate more and more. In this way, it has been able to attack the small- to medium-sized enterprise market effectively and decisively, creating positive incentives for both sides. Because 68 percent of SAP's revenues come from its software (31 percent is consulting and training), it has the luxury of taking this strategy. Its most recent endeavor, mySAP.com, which is a joint venture with Fujitsu, Siemens AG, and numerous smaller companies, is an attempt to further leverage its client-server architecture to create a point-to-point Internet buying and selling solution among all of its users. The idea is to create a digital marketplace to enable intercompany relationships for buying, selling, and communicating in an open electronic business-to-business hub. Continuing innovation in a rapidly growing market requires numerous alliances with other complementary partners and host governments alike.

Case Study 2:
Merant Micro Focus—Global Integration and Industry Consolidation

Throughout the 1980s, the original Micro Focus had few challengers in the specialized field of software development tools using the Common Business-Oriented Language (COBOL) development language, allowing it to become the technological darling of the UK stock market in the 1980s. But with growing PC and network platforms emerging, there were nagging questions beginning in the late 1980s as to whether this competitive advantage would enable it to sustain both its growth and profitability. Furthermore, in spite of continual revision, it remained a primitive and awkward language, seemingly ripe for replacement by more modern languages.

But resistance to change in the computer industry and the huge investment computer users had made in COBOL and software specialists skilled in COBOL meant that large corporations were willing to work with their existing environment. As late as 1991, there were over one million programmers worldwide, all of whom could benefit from Micro Focus's special tools to make writing COBOL programs easier. Micro Focus had just begun to penetrate this vast market in the late 1980s. The fact that it continually sells more of its products (53 percent) in the United States than in either Europe (36 percent) or Japan (11 percent) meant that it had to behave like a global company. Accordingly, it capitalizes research and development expenditure, a practice generally shunned by European software firms but common among U.S. companies, and it does not pay dividends. By 1991, close to 25 percent of its stock was held by U.S. investors in the form of ADRs and two years later, it was quoted on the U.S. NASDAQ electronic stock exchange.

In order to compensate for the lack of an effective venture capital market and misunderstanding of the software market at home, Micro Focus used a sustained high-profile presence in the U.S. market to get informed U.S. investors on board to fund its vision of growth. But what gives U.S. investors such confidence in a small UK software house whose stock-in-trade is unrivalled expertise in a thirty year-old computer language, which by most measures should be dead and buried by now? The network externalities factors discussed earlier are particularly relevant to Micro Focus. With the largest installed base of COBOL clients in the world, it made sense to focus on the U.S. market. In addition, since the dominant hardware producers were major U.S. companies, it was a stable investment to commit to the creation of a large Micro Focus's U.S. subsidiary to ensure both user and producer interaction to meet market needs. By the early 1990s, Micro Focus had some 440 staff in Palo Alto, Philadelphia, Chicago, and Los Angeles and has licensed its COBOL products to every sizeable computer manufacturer in the world. By the third quarter of 2000, the staff size has tripled and their operations branched out to over fifty countries, including subsidiaries in Japan, South Korea, and Singapore.

Envisioning the virtual ownership of the microcomputer COBOL market, building capabilities for speed, surprise, market positioning, and simultaneous market thrusts enables Micro Focus to stay one step ahead of the game. In 1988, Micro Focus launched a product called Micro Focus COBOL/2 Workbench that enabled users to work on their mainframe program with microcomputers, the new global networking standards. Because it was much easier to work with a desktop computer than an unwieldy mainframe, this innovation saved money and enhanced productivity. Indeed the success of the group's products has given a whole new lease on life to COBOL, a computer language invented in the 1950s. The product's appeal is irresistible even in a recessionary climate, and sales to corporations all around the world have grown explosively. About two-thirds of group turnover is now generated by Workbench, and profits have grown from £1.6 million ($2.4 million) in 1988 to an expected £12–14 million ($18–21 million) in 1998.

In addition, this Newbury, Berkshire-based company has a demonstrated track record of successful adaptation. For example, one of Micro Focus's most important achievements to date was the 1990 announcement of an extensive strategic marketing and development agreement with IBM. By becoming an IBM business partner, Micro Focus not only gets access to working with the world's largest computer manufacturer to create better methods of developing business software, but it also gets access to tap into the top tier, worldwide IBM distribution and marketing network for its products.

Since Micro Focus's version of COBOL, the world's most-used business computer language, is the official IBM language for use on computer workstations, this ensured a steady revenue stream from which to leverage into other markets. Even Microsoft was still using Micro Focus's version of COBOL until it fully committed to Windows NT in the early 1990s.

Through leveraging its competitive advantages in COBOL-based products to become the world leader in key market niches, it was able to remain innovative in finding new ways to satisfy customers with old systems by allowing them to take advantage of recent technological developments. Through an aggressive global strategy centered on satisfying the large installed base of COBOL users in the U.S. market, Micro Focus is well positioned potentially to do well in Asia as its partners expand. The core strategy is to remain an innovative player capable of designing products to link old and new systems, thereby ensuring that it will be able to expand beyond Japan as new markets open up. In fact, its Workbench products have proven to be equally satisfying for microcomputer link-ups to Japanese mainframe and minicomputer systems as well. As a result, the company has managed to stay competitive, serving the U.S., European, and then Japanese markets respectively as entrees into developing Asian countries.

But in the short-term, Micro Focus was focusing on consolidating its U.S. market share, the core of its operations. In order to stay on top of this rapidly evolving market and realizing its current products would have finite market niches, the firm sought to become a developer of integrated enterprise applications. To become a global e-business software solution provider would require updating internal capabilities through various alliances and acquisitions. In 1998, in an effort to build on its previous success as a COBOL tools supplier, Micro Focus bought Intersolv to create a company capable of providing a complete tool set for IT departments. Both companies have profited from the high demand for tools to fix the Y2K problem and realize that this is only a short term market niche. The combined venture competes with leading tools vendors, such as IBM and Sterling Software, which bought out UK-based Synon in 1999 (see next case study). This was the first step toward creating the basis for serving the booming Internet market as an e-business software solution provider.

Intersolv specializes in decentralized computer systems and developing software for the expanding business Internet market. In the first quarter of its financial year to July 31, 1998, Intersolv's profits tripled over the previous year to $2.1 million (£1.3 million). For Micro Focus, which is listed on both the Stock Exchange and on the U.S. technology-dominated exchange NASDAQ, this was the next logical step to becoming one of the largest concerns exclusively focusing on developing software tools for large corporations that desire a single supplier providing both tools and services for their computers. In addition to assembling a computer package with special manufacturing tools, large corporations also need development tools for creating software applications, as well as the ability to provide services for putting a final package together and testing it. Intersolv gave them the capability to integrate the latter into their now comprehensive operations.

Since the companies had been doing business together since 1988, the merger was a natural fit. According to Micro Focus chief executive Martin Waters, who headed the combined company, "what drove this [deal] was being responsive to our customers. Neither company could offer a total solution before."[73] Since IT applications are often written in various languages and run on a variety of computer platforms around the world, Waters believes that this alliance would now enable the two companies to generate more market share through the capacity to eliminate language barriers between computer systems. The new entity was renamed Merant Micro Focus in an effort to re-brand themselves as integrated enterprise software suppliers for multiple platforms in a global e-business environment.

In the fiscal year ending April 30, 2000, revenues reached $365.4 million (despite the high expense of acquisitions), experiencing growth at a rate of 8–10 percent a year since fiscal year 1998. The stunning success of Micro Focus provides a useful model for other European software companies. It offers a three-step entry strategy into Asia. First, it involves building on existing core competencies while simultaneously acquiring new capabilities to offer a standardized software product solution to their primary European clients. Second, the majority of their acquisition efforts were then focused on the most cutting-edge technological market, the United States, which possessed the dominant market share for their first global software product. Third, as the Internet and new distributing computing environments spread to Asia, Micro Focus would be well-positioned to partner with new emerging market opportunities in the fastest growing region, Asia; and the biggest market segment, the integrated enterprise solutions niche.

Typical of their European counterparts, Micro Focus was founded in 1976 to design software for proprietary hardware makers such as Siemens, Bull, and Alcatel using COBOL. As their growth slowed, so did Micro Focus. But because they were committed to a vision of developing tools for open systems in their research and development, they were well-positioned to take advantage of the short-term demands generated by the millennium bug and the need to equip computers to handle transactions in euros once the single European currency was launched. In the words of one analyst, " . . . reservations about Micro Focus were based on the fact that the company was too reliant on COBOL as the main computer language for the mainframe environment; its growth was confined to the Y2K problem; and the level of services was not sufficient."[74]

Buying Intersolv addressed all three of these issues. By broadening the range of products and services, the deal gave Micro Focus a growth market beyond 2000. But more importantly, Intersolv, which operates in different computer languages, took Micro Focus into a service area from which it has been absent. Although service and maintenance provides 45 percent of the combined group revenues, which on a pro-forma basis would have totaled an estimated $350 million for 1997, the capacity to sell a combined software and services package to the booming Asian markets is now a reality. Currently two-thirds of total revenue comes from North America, but a strategy based on building long-term alliance partners—following the SAP lead—appears to be in the works.

Although Intersolv did not come cheap—Micro Focus paid almost three times its sales for the company (52 percent to the expected combined revenues)—investors signaled their approval in Intersolv trading on the NASDAQ as a cheap way into Micro Focus. In fact, Waters said the deal would lift earnings per share and increase revenues in the first full year. As a result, in building capabilities for speed and surprise, Micro Focus has already announced plans for further acquisitions of similar technologies. By buying the privately held XDB Systems of the United States, a provider of development, maintenance, and connectivity tools for the DB2 database standard, Micro Focus makes a credible commitment to aggressively pursue several markets at the same time. Holders of XDB shares will receive Micro Focus shares based on a formula that values XDB at $13.4 million with a special allocation of $3.1 million for certain assets of XDB. Because of XDB's dismal performance in 1998, which lost $3.2 million on sales of $10.1 million and had assets of $13.1 million, Micro Focus attained valuable complementary assets. Furthermore, Micro Focus acquired Proximity Software for $4.1 million, to be satisfied by the issue of 120,000 new shares in late January 1999. The acquisition was described as a pooling of interest since Proximity technology already uses Micro Focus's products.

Since then, Merant has become a major player in the enterprise application development market segment, providing the products, people, and processes to help businesses adapt to the rapidly evolving IT innovations. Their main product, the MERANT Egility framework helps organizations adapt their enterprise applications for the changing technology and business requirements of the e-business environment, manage the software application development process, and provide integrated data connectivity across the enterprise, from the mainframe to the Internet. By the end of 2000, Merant had become a global organization with approximately $370 million in annual revenues and nearly two thousand employees, with over five hundred technology partners and more than five million licenses at over thirty-five thousand customer sites (including the entire Fortune 100 and the majority of the Global 500).

Case Study 3:
Synon—From Early Adopter to Acquisition

In many respects, Synon of North London is a classic entrepreneurial story. Early on, this small company, founded in 1983, saw a need for making application programs easier to develop. Synon was convinced that the U.S.-led PC revolution and cost-conscious European spending habits translated into a global market for applying engineering techniques to develop software making tools for various platforms. As a result, Synon became one of the pioneers in the design of computer-aided design engineering (CASE) tools. Although initially targeting tools for proprietary platforms for European markets, plans to develop a global product were assisted by UK science and technology policies as well as European initiatives. Through the provision of favorable tax policies from the former and development funds from the latter, this small company became the software darling of the country, winning the prestigious Queen's Award in 1991. Later in that same year, their new CASE tool, Synon II, designed specifically for the mid-range IBM AS/400, earned the firm recognition from IBM as "business partners"—the first UK company to do so. Buying an initial stake of 10 percent in the company, IBM gradually increased its shares as revenue streams drastically increased throughout the next five years.

Synon's global strategy was shaped not only by direct association with IBM, but also by the hiring in 1993 of a high-level executive from IBM. Following the U.S. model, efforts were made to expand capabilities to deliver solutions quickly and effectively. A CASE-centric view of the world was envisioned and an aggressive pursuit of market share required a simultaneous strategy of acquiring smaller software firms to consolidate their temporary market niche lead, setting up strategic alliances with multiple partners in various countries, and building the capacity to serve multiple platforms. Accordingly, three small U.S. companies were bought out between 1991–95 and alliances with Microsoft and CCCL were created to launch new case tools for Windows NT where Malaysia was one of the first intended markets.[75] By the end of 1996, over 500 new clients in Malaysia and the United States were using Synon CASE tools. However, Synon's successful run finally ended when Sterling Software, Inc., agreed to acquire the privately held Synon, Inc., for $79 million in stock. Sterling Software, based in Dallas, Texas, provides software and services for applications management, systems management, and federal systems. It had revenue of $489 million in the year that ended on September 30, 1997. Synon, which had already "turned American" for the majority of its operations, was based in Larkspur, California, and provided AS/400 and AS/390 application development products. It had revenue of $79 million in 1997 at the time of its acquisition. By the middle of 2000, the legacy of Synon was further integrated into an even larger corporation, Computer Associates, through the creation of CA/Sterling. With Synon CASE tools, Sterling's portal technology, storage and network management tools, and data warehousing expertise, the new entity now promises to deliver a comprehensive platform for global e-business application development, intelligence, security, customer relationship management, and dynamic customization.

V. Conclusion

The initial discussion began with a story of a dying, customized, segmented, and craft-like European software sector. It then went on to discuss the dynamics behind the rapidly evolving global software industry as a benchmark for the state of the current European industry. As discussed in chapter 1, firms do not exist in a vacuum and their institutional history affects how they attempt to adapt to new conditions.[76] The successful case studies above leveraged a specific market niche to attach simultaneously and sequentially multiple platform markets across the globe. Focusing on the new, emerging economies of scale, they built organizations designed for speed and surprise, aggressively pursuing international strategic alliances and various types of interfirm arrangements to minimize time to market. In this marketplace characterized by self-reinforcing network externalities, market share and an established base of installed users is crucial to surviving in a rapidly evolving software industry.

As discussed in earlier sections and in the other chapters of this volume, the primary drivers of successful entry into the Asian systems, services, and packages software markets are: (1) an in-depth knowledge of users' needs that are normally developed through long-term relationships with customers; (2)

extensive marketing and distribution networks with localization and support capabilities; (3) long-term access to capital with government assistance and cooperation; (4) reputation, which is particularly crucial to potential clients; (5) differentiation from competitor products; (6) timely and consistent product offerings and upgrades that lower their clients' overall need for IT; (7) strong internal market intelligence (R&D) capabilities to take advantage of technological innovations as they become commercially viable; (8) interaction with policymakers to ensure that the nonmarket environment continues to provide favorable incentives to help amortize expensive R&D costs; and (9) the establishment of long-term relationships with host governments and firms alike. Many of these obstacles were overcome through international strategic alliances and direct foreign investment. But given the increasingly massive costs of research and development in an industry of rapid technological change and shortening product cycles, pan-European, national, and local initiatives help firms amortize the transaction costs, avoid wasted duplications of efforts through coordinated standards bodies, create a cooperative environment necessary to mitigate high risks and uncertainty, and mutually envision a global information superhighway.

The three global packaged software suppliers discussed in these case studies have found that the market moves rapidly and survival means attacking multiple fronts simultaneously. As was the case with Synon, failure to acquire new capabilities and quickly bring new software products to market could have a devastating effect on a firm within a short time period. The path of Synon does not appear to be atypical of the numerous, small European ISVs, which lacked an aggressive global strategy that required complex strategic alliances to remain viable. As producers that were used to providing for a steady base of local clients (usually MNCs and governments) in similar industrial niches, it is not surprising that many opted out of attaining the apparently larger size necessary to remain viable in an industrial sector that experienced rapid global consolidation throughout the 1990s. Arguably then, it is not yet clear that Merant will survive the next wave of consolidation and suffer a similar fate as Synon (and Intersolv).

But what is clear is that all of the firms in the case studies initially linked their global strategic vision with the skillful and creative leveraging of regional advantages, while building upon their path-dependent organizational resource capabilities to leverage specific market niches into an integrated strategy. Given the rapidly increasing size of Asian markets and intimate connection with telecommunications, entry into Asia will enable new global players to capture the potentially self-sustaining benefits associated with the network externality dynamics of the industry. To do otherwise and let the

United States continue to dominate these markets could result in further decline in competitiveness across the board. However, the success of SAP, Micro Focus, and Synon provide useful case studies with mixed results that offer much hope and many lessons for other European software vendors.

VI. Appendix:
Methodological Notes on Case Selection

Why "Traded" Software?

For practical and methodological concerns related to data collection, the focus of this chapter is primarily on European firms in the "traded" software sector with secondary emphasis on the "packaged" industry segment. "Traded" software refers to software that is produced by one firm for sale to another.[77] "Packaged" refers to standardized software that offers general application solutions that are not "customized" for a particular end user, industry, or business. While all packaged software is traded, customized software may also be traded. Unlike packaged, custom software is normally sold in conjunction with broader "computer services."[78] Focusing on traded software intended for sale is practical since it establishes boundaries for the unit of analysis. Because a great deal of software development is carried out within user firms, it is not surprising that internationally comparable data on sales of different types of software do not exist.[79] However, data for traded packaged and custom software, the dominant type of software produced by European firms, does exist.[80]

Emphasis is given to "packaged" software because it is the fastest growing software segment in the world (see table 5.7). Furthermore, as increasingly sophisticated packaged products (including development tools) become available in line with rapidly growing PC, LAN, WAN, and workstation environments, the market share of purely customized solutions is steadily declining (see table 5.8). The key words now are connectivity, compatibility, and commonality (standard graphical user interfaces (GUI). Thus, it is not surprising that the dominant global players of the industry are "packaged" software firms, primarily from the largest market, the U.S. However, since few European players are global packaged players—and most are still years away from being able to offer competitive standard solutions, I will focus on the European software firms that satisfy two conditions: (1) have some degree (or potential) of export-orientation toward Asia; and (2) have established a niche in some type of traded software product. This allows for a much more manageable data set of a handful of firms as opposed to the over twenty thousand independent software vendors and services firms that exist in Europe.

The focus on traded software in the enterprise solutions segment also makes sense for methodological reasons as well. First, the rapid diffusion of both cheaper workstations and microcomputers has generated widespread opportunities for creative software solutions in a broad range of highly specialized applications. As a result, at least four groups of firms are active in the traded software sector: (1) computer hardware producers; (2) independent software vendors; (3) independent computer service firms, including value-added resellers; and (4) IT consulting firms. Furthermore, telecommunications deregulation, the advent of the Internet, wireless communications, and other technological developments have resulted in even tighter coupling between the computer software business and "computer services/consulting" sector.[81] For example, Cap Gemini, one of Europe's largest remaining computer services/consulting companies, does significant business in Asia (close to $70–80 million in 1997) that involves a suite of offerings of standard software products that also involve process and service operations as well. Its "telecom in a box" solution in fact involves a vast array of products, processes, and services more akin to a standard customized solution of which software is one integrated part.

Although determining boundaries in the traded software sector does not escape these problems, the focus on firms that are primarily producers of traded software for sale to a global market allows analysts to track a specific commodity. Disaggregating the software creation value of a consulting or computer services firm that has a general IT contract for a large organization would be difficult to discern as well as compare. However, tracking the contribution of the packaged software component from services such as installation, systems integration, and maintenance is a more manageable task than disaggregating a customized solution. In addition, only firms that have a packaged software product are likely to penetrate the Asian software market to any significant degree.

Notes

1. Arthur, W. B. (1989), pp.116–131; David (1993).
2. Steinmueller (1996). A few of the major factors include: a favorable defense "spinoff" policy, distinct market conditions, a long history of business-university collaboration, active venture capital markets, and a strong entrepreneurial culture. But this list is by no means exhaustive.

3. PC refers to personal computer, not necessarily the IBM-compatible PC platform. Apple's Macintosh computers are also considered PCs as used in this essay.

4. Ascertaining why European producers have evolved into this market niche is beyond the scope of this chapter. But it is clear that market incentives by the late 1980s led many firms to seek standardized solutions for similar types of large-scale organizations.

5. Hoch, et al. (2000).

6. See Appendix for methodological concerns related to case selection.

7. National Research Council (1992).

8. This classification was developed by the International Data Corporation (IDC), who make the distinction between software that is traded as a commodity and software that is either embedded in a product or only a part of the overall professional service being purchased. Other IT analysts such as the Gartner Group and Forrester Research tend to follow this classification.

9. Ovum Consultany (1997, 1998).

10. Ibid.

11. See chapter 1 by Aggarwal in this volume.

12. I justify this focus for three main reasons. First, it is the only European software firm that is a worldwide household name. Second, there is an enormous amount of information on the firm, which has recently switched to U.S. general accounting principles. Third, as an early entrant in the industry, it has demonstrated robust growth and the ability to change with the times.

13. David, Paul A. (1985), Arthur (1989), D'Aveni (1994), and Varian and Shapiro (1999).

14. IDC (2000).

15. IDC (2000). Packaged segments grew between 16.4–17.4 percent in 1999, over the 14 percent overall growth rate.

16. See Arthur (1989), David (1993), and Mowery (1996) for a discussion on network externalities.

17. *The Economist,* "The Software Industry," May 25, 1997. Many European software entrepreneurs ended up locating directly in the United States as their only chance for success.

18. Cusumano and Yoffie (1998).

19. Varian and Shapiro (1999).

20. Cusumano and Yoffie (1998); and Varian and Shapiro (1999).

21. The pre-Asian crisis totals were $3.9 billion, which made it the highest ranked region for losses in 1997.

22. Business Software Alliance (1998, 1999, 2000).

23. Bresnahan and Trajtenberg (1995), pp. 83–108. The authors argue that a market economy may not be suitable for the development of general purpose technologies like computers and software packages because the production

of these technologies is subject to externalities associated with severe coordination problems between producers and users.

24. Flamm (1988); and Teece and Grindley (1997).
25. In fact, the next major battle is currently occurring in the home appliance network sector that promises to enable one to interconnect all of one's home appliances into a single integrated interface.
26. ESPRIT (1997), p. 1.
27. See *The Economist,* November 12, 1994, pp. 77–78. As mentioned earlier, high research and development costs (R&D), high training costs, increasing returns to scale, rapid technological change, extreme uncertainty, and a craft-like nature all characterize the software industry. See Saxenian (1994).
28. Miles and Snow (1986), Saxenian (1994).
29. Gibbs, W. W. (1997).
30. Gibbs (1997), pp. 72–81. According to Gibbs, for every six new large-scale software systems put into operation, two others are cancelled. In addition, the average software development project overshoots its schedule by half with the largest projects performing on the worst end of the distribution.
31. ESPRIT (1997), p. 1.
32. Maxwell, Van Wassenhove, and Dutta (1998), Dutta, Van Wassenhove and Kulandaiswamy (1998). According to the authors, there were only two published studies of the state of Europe and success of European-wide initiatives that describe the real experiences of firms in the last five years.
33. According to Venture Economics (a research organization), only 7 percent of European investment banks money went into computer-related companies, compared with 24 percent in the United States (a value of $4.4 billion to $10 billion in 1993).
34. See Dutta, et al. (1998), and Torrisi (1998).
35. *The Economist,* November 12, 1994.
36. Joly, Kluge, and Stein (1994), pp. 33–38.
37. Saxenian (1994).
38. In part, many strategic experts argue that "giving away" one's product can be an essential part of an aggressive market share strategy. See Varian and Shapiro (1998).
39. Teece (1986).
40. The Computer Science and Technology Board (1993).
41. Snow, Lipnack, and Stamps (1999), pp. 615–630.
42. Torrisi (1998).
43. See Torrisi (1998), p. 37, and SAP's website at www.sap.com.
44. Hoch, et al. (2000), p. 36.
45. OECD (1985).
46. Torrisi (1998), p. 91.
47. OECD (2000).

48. According to *The Economist* (1996), 70 percent of the industrial revenues were from software products that did not exist two years ago. Furthermore, according to Hoch, et al. (2000), 60 percent of startups in the ISV business go bankrupt in three years.
49. IDC (2000), Moschella (1997), and Malerba and Torrisi (1996).
50. In fact, this is the mission objective behind the original conception of SAP. See Maeissner (1997), p. 21.
51. National Research Council (1992).
52. D'Aveni (1994).
53. Dedrick and Kramer (1998).
54. One exception Bandinelli, Fuggera, Lavazza, Loi, and Picco (1995), pp. 440–454; and Torrisi (1998).
55. See chapter 1 by Aggarwal in this volume, and Barron (2000), for more detail.
56. Barron (2000), chapter 1.
57. Malerba and Torrisi (1996).
58. Borrus and Zysman (1997, 1998).
59. Hoch, et al. (2000).
60. Vernon (1985).
61. Malerba and Torrisi (1996).
62. IDC (1994).
63. *The Economist,* "Telecommunications Survey," September 13, 1997.
64. Vernon (1985).
65. Anderson and Gatignon (1986).
66. Gomes-Casseres (1992).
67. Ibid.
68. *Fortune,* December 7, 1998, p. 103.
69. According to SAP's website at www.sap.com/main.
70. D'Aveni (1994).
71. D'Aveni (1994), p. 244.
72. *Fortune,* December 7, 1998, p. 102.
73. Ibid.
74. Ovum Consultancy (1998), p. 6.
75. *Financial Times,* March 15, 1996, p. 4.
76. See chapter 1 by Aggarwal in this volume.
77. Mowery (1996), p. 5.
78. However, given the increasing complexity of some packaged solutions intended for large firms or organizations, the need for related computer services is growing at a rate that by some estimates, exceeds that of the packaged market itself. See USDOC (1998).
79. Steinmuller (1996).
80. The OECD, the International Data Corporation (IDC), the U.S. Department of Commerce, the National Trade Development Bank, the European Commission, and many private consulting companies that specialize in IT collect

data in both packaged and customized software. This does not imply that their categories are neither identical nor comparable without qualification.

81. Even the OECD and IDC are not in agreement as to what constitutes services (e.g., systems integration).

References

Anderson, Erin and Hubert Gatignon (1986). *The Multinational Corporation's Degree of Control Over Foreign Subsidiaries: An Empirical Test of a Transaction Cost Explanation*. Cambridge, MA: Marketing Science Institute.

Arthur, W. Brian (1989). "Competing Technologies, Increasing Returns and Lock-in by Historical Events." *Economic Journal,* pp. 116–131.

Bandinelli, Salvatore, A. Fuggera, L. Lavazza, M. Loi, and G. Picco (1995). "Modeling and Improving an Industrial Software Process." IEEE Translation, *Software Engineering* 21(5), May, pp. 440–454.

Baron, David P. (2000). *Business and its Environment,* 3rd Edition. Upper Saddle River, NJ: Prentice Hall.

Borrus, Michael and John Zysman (1997). "You Don't Have To Be a Giant: How the Changing Terms of Competition in Global Markets Are Creating New Possibilities for Danish Companies." BRIE Working Paper no. 96A. Berkeley: Berkeley Roundtable on the International Economy.

———(1998). *Globalization with Borders: The Rise of Wintelism as the Future of Industrial Competition.* BRIE Working Paper no. 96B.

Bresnahan, Timothy F. and Manuel Trajtenberg (1995). "General Purpose Technologies: Engines of Growth?" *Journal of Econometrics,* pp. 83–108.

Business Software Alliance (1998, 1999, 2000). *World Software Piracy Report.* Washington D.C.: BSA.

Cusumano, Michael A. and David B. Yoffie (1998). *Competing on Internet Time: Lessons From Netscape and its Battle With Microsoft.* New York: The Free Press.

D'Aveni, Richard (1994). *Hypercompetition: Managing the Dynamics of Strategic Management.* New York: The Free Press.

David, Paul A. (1985). "CLIO and the Economics of QWERTY." *American Economic Review:* 332–337.

———(1993). "Path-Dependence and Predictability in Dynamic Systems with Local Network Externalities: A Paradigm for Historical Economics." In *Technology and the Wealth of Nations,* edited by D. Foray and C. Freeman. London: Pinter Publishers.

Dedrick, Jason and Kenneth Kramer (1998). *Asia's Computer Challenge.* Oxford: Oxford University Press.

Dutta, Soumitra, Luk N. Van Wassenhove, and Selvan Kulandaiswamy (1998). "Benchmarking European software management practices," *Communications of the ACM.* 41(6) June, pp. 77–86.

ESPRIT (1997). "Software Best Practices (ESSI)." at www.cordis.lu/esprit/src/ essi.htm.

Flamm, Kenneth (1988). *Creating the Computer.* Washington D.C.: The Brookings Institution.

Gibbs, W. W. (1997) "Software's Chronic Crisis." *Scientific American* (Int. Ed.) 271(3), August, pp.72–81.

Gomes-Casseres, B. (1992). "International Trade, Competition, and Alliances in the Computer Industry." Working paper 92–044. Boston: Division of Research, Harvard Business School Press.

Hoch, Detlev J. et al. (2000). *Secrets of Software Success.* Boston: Harvard Business School Press.

International Data Corporation (IDC) (1994). *Packaged Software 1993.* February 1.

IDC (1997). *Asian Computer Markets.*

IDC (2000). *Packaged Software 1999.*

International Planning and Research Corporation (2000). *1999 Global Piracy Report.* Washington, DC: Business Software Alliance and Software and Information Industry.

Joly, Herbert, Jurgen Kluge, and Lothar Stein (1994). "Europe's Structural Weakness." *The McKinsey Quarterly* 1, pp. 33–38.

Maeissner, Gerd (1997). *SAP—die Hemiliche Software-Macht (SAP—The Secret Software Power).* Hamburg, Germany: Hoffman and Campe.

Malerba, Franco and Salvatore Torrisi (1996). "The Dynamics of Market Structure and Innovation in the Western European Software Industry." In *The International Computer Software Industry: A Comparative Study of Industry Evolution and Structure,* edited by David Mowery. Oxford: Oxford University Press.

Maxwell, Katrina, Luk Van Wassenhove, and Soumitra Dutta (1998). "Performance Evaluation of General and Company Specific Models in Software Development Effort Estimation." *Management Science,* 45(6), pp. 787–803.

McKinsey & Company (1996). *The 1996 Report on the Computer Industry.* New York: McKinsey & Company.

Miles, Raymond and Charles Snow (1986). "Organizations: New Concepts for New Forms." *California Management Review* 28(3).

Moschella, David C. (1997). *Waves of Power.* New York: AMACOM.

Mowery, David (1996). *The International Computer Software Industry: A Comparative Study of Industry Evolution and Structure.* Oxford: Oxford University Press.

National Research Council, The Computer Science and Telecommunications Board (1992). *Keeping the U.S. Computer Industry Competitive: Systems Integration.* Washington, D.C.: National Academy Press.

OECD (1985). *The Software Industry.* Paris: OECD.

OECD (1997). *Information Technology.* Paris: OECD.

OECD (2000). *OECD Information Technology Outlook.* Geneva: OECD.

Ovum Consultancy. *IT in the EU* (1997–98).

Saxenian, Annalee (1994). *Regional Advantage: Culture and Competition in Silicon Valley and Route 128.* Cambridge, MA: Harvard University Press.

Siwek, Stephen E. and Harold W. Furchtgott-Roth (1993). *International Trade in Computer Software.* Westport, CT: Quorum Books.

Snow, Charles, Jessica Lipnack, and Jeffrey Stamps (1999). "The Virtual Organization: Promises and Payoffs, Large and Small." *Journal of Organizational Behavior,* pp. 615–630.

Steinmueller, W. Edmund (1996). "The U.S. Software Industry: An Analysis and Interpretive History," In *The International Computer Software Industry: A Comparable Study of Industry Evolution and Structure,* edited by David Mowery. Oxford: Oxford University Press.

Teece, David (1986). "Capturing Value from Technological Innovation: Integration, Strategic Partnership, and Licensing Decisions." International Business Working Paper No. 1 B-6. Berkeley, CA: Haas School of Business.

Teece, David and Peter Grindley (1997). "Managing Intellectual Capital: Licensing and Cross-Licensing in Semiconductors and Electronics." *California Management Review* 39 (2).

Torrisi, Salvatore (1998). *Industrial Organization and Innovation: An International Study of the Software Industry.* Cheltenham, UK: Edward Elgar.

United States Department of Commerce (USDOC) (1998). *World Computer/IT Software/Services Best Markets Report,* November 13. Washington D.C.: Department of Commerce.

Varian, Hal R. and Carl Shapiro (1999). *Information Rules: A Strategic Guide to the Network Economy.* Boston: Harvard Business School Press.

Vernon, Raymond (1985). *Exploring the Global Economy: Emerging Issues in Trade and Investment.* Lanham, MD: University Press of America.

Chapter 6

The Fast Lane to Asia:
European Auto Firms in China

Nick Biziouras and Beverly Crawford

I. Introduction

The decade of 1985–1995 was an important watershed in the history of the international automobile industry. World demand for automobiles had stagnated. Declining international competitiveness had thrown North American and European automobile manufacturers into labor turmoil. Overcapacity threatened home markets that had already achieved predictable and mature growth rates, resulting in a glut of excess manufacturing capacity, now estimated at about forty unneeded assembly plants world-wide. Japanese-U.S. and Japanese-European Union (EU) trade relations were increasingly strained, as Japanese automobile manufacturers penetrated western markets, while carefully protecting their home turf.

While these problems festered, the Asian auto market was exploding. With the exception of Japan, the remaining last frontier for market penetration appeared to be Asia: economic growth rates were high throughout the region; a middle class with a significant disposable income was emerging; and few people owned cars. But European and American firms faced formidable Japanese competition: Japanese manufacturers had built an important presence in Asia through decades of market penetration in sales and the location of manufacturing facilities.

Rapid economic development in China in the 1980s, however, provided the promise of large market share to all those who could establish a presence there; Japanese manufacturers had never gained a toehold in the Chinese market. Domestic producers had long manufactured vehicles for government purposes, producing limited models with standard technology. Thus, in the absence of Japanese competition, China's 1.3 billion people and rapidly developing economy appeared to be one of the few great growth plays left for European producers. By the mid-1980s, the rapid rise in per capita income held the promise of opening new market segments for a wide range of models, including compacts, light trucks, minivans, mid-sized cars, and even luxury vehicles. As personal income rose even more rapidly in the 1990s, analysts saw on the horizon an increasing demand for large, chauffeur-driven sedans for government officials and for the expanding number of foreign businesspeople and entrepreneurs beginning to flood China. The market was wide open, and, betting on growing demand, European producers rushed to grab market share. By 1999, Volkswagen (VW), with 55 percent of all passenger auto sales, dominated the Chinese market; Peugeot and Volvo, however, pulled out early, after initial failures resulted in huge losses.

This chapter analyzes the causes of Volkswagen's success and Peugeot's failure in penetrating the Asian market by locating production facilities in China. Despite the greater relative persistence of central planning, protected market, and a lack of infrastructure, China displays two factors central to any analysis of market penetration in Asia: First and often ignored by the conventional focus on state planning, China possesses all of the characteristics of a "developmental state" common to many Asian countries.[1] Thus, nonmarket political influences provide important constraints and opportunities for market entry and control. Second, China's growing economy provides even more access points and incentives for market entry than the economies of other Asian developing countries. This growing economy creates the possibility for huge economies of scale for the manufacture and distribution of industrial products for foreign multinational firms, and it also provides those firms with an expanding, consumption-oriented middle class. Demand for autos is arguably more pronounced in China than in other parts of developing Asia.

In this context, we provide a comparative case study to investigate the interaction of market and nonmarket strategies in determining the success or failure of market penetration of the automobile sector. The experience of Peugeot and Volkswagen in China suggests that both strategies can be fruitfully investigated. Both companies shared the following motivation and action: (1) they set up production facilities within a year of each other; (2)

both attempted to escape high labor costs in Europe by relocating production facilities to areas with low-cost labor; (3) both were anxious to expand market share under the pressure of stagnating global demand; (4) both expressed a commitment to establish a production base in China in order to penetrate the rapidly growing Southeast Asian market; and (5) both were among the first auto firms to be allowed access and entrance in the Chinese market. Further, each firm's home government took measures to promote auto exports and investments abroad, and in both cases there was a long-standing tradition of close government-firm collaboration to protect the national auto industry from global competition. Finally, both firms benefited equally from EU efforts to exploit overseas investment opportunities and the increased liberalization of emerging economies, especially China and India. Indeed, the EU managed to avoid any connections between the human rights violations in Thailand and the 1996 Asia-Europe Meeting (ASEM), which focused on the creation of closer economic and trade ties.[2] Despite these important common incentives and constraints, VW succeeded in gaining a foothold in the Chinese market, and Peugeot failed. By the end of 1999, VW had a 55 percent share in the Chinese market, and Peugeot had sold its production facilities in China to Honda. Why?

To address this puzzle, we begin with a description of the regional context and chief factors that shaped the investment strategies of Western auto firms in Asia over the last two decades. This is followed by a comparative study analyzing the factors that contributed to both success and failure. In so doing, we describe specific market and nonmarket forces that pose both threats and opportunities and thus frame the conditions for the competitive success of European firms.

II. Positional Analysis

Geographical Orientation

One can quickly see why the Chinese market has gained importance: China posted an average annual Gross Domestic Product (GDP) growth rate of 10.2 percent in the 1980–1990 period, which was significantly higher than that of India (5.8 percent), Indonesia (6.1 percent), Japan (4.0 percent), South Korea (9.5 percent), and Malaysia (5.2 percent). This dramatic growth did not decelerate in the early part of the 1990s. Between 1990–1997, China averaged an annual GDP growth rate of 11.9 percent while India grew by 5.9 percent, Indonesia by 7.5 percent, Japan by 1.4 percent, South Korea by 7.2 percent, and Malaysia by 8.7 percent.

This rapid economic growth was mirrored in the annual growth of per capita private consumption. Between 1980–1996, China grew by an average annual per capita rate of 4.5 percent; in comparison India grew by 1.6 percent, Indonesia by 2.8 percent, Japan by 2.9 percent, and Malaysia by 1.7 percent. In this sense, what Asia, and China in particular, offered was a chance for expansion that most other markets could no longer provide.

Although sectoral statistics rarely speak for themselves, the growth potential of the Asian markets for automobiles could hardly have been lost on any of the interested parties, whether they were international automobile manufacturers, host national governments, or trade unions in the advanced industrial markets. In terms of auto registration, Asia witnessed an average rise of 46.3 percent between 1981 and 1988, with rates rising by 31.8 percent in Japan, 204.6 percent in Taiwan, 210.8 percent in South Korea and 343.3 percent in China. Even if the important South Korean and Japanese industries are excluded, Asia's share of global auto production increased from 1 to 3 percent.[3] The increases in registration and production continued well into the 1988–1991 period, with auto production increasing by a factor of 4 in South Korea, by 1.5 in Taiwan, and by 2.6 in Thailand. Similarly, auto registration increased by a factor of 4.3 in South Korea, by 3.2 in Taiwan, and by 3.5 in Thailand.[4] In the 1992–2000 period, the forecasts for annual compound growth rates for new car sales were between 21.9 percent for China, 10 percent for Malaysia, and 4.8 percent for Indonesia, significantly higher than the world average of 2.4 percent.[5] Although dampened by the Asian financial crisis of 1997, growth rates continue to exceed those of most other developing countries. As Vaughn Koshkarian, the president of Ford China, optimistically stated at the onset of the 1997 financial crisis, "by 2010 China will have four vehicles per one hundred people and a market volume of between five and six million vehicles, the fourth largest market in the world after North America, Europe and Japan. [Additionally,] by 2010, after substantial consolidation, this automotive industry will have a highly educated, skilled and industrious workforce. In essence, China will have everything necessary to become a primary manufacturing nation in Asia."[6] Robert Buscelhofer, a member of Volkswagen's car management board, underlined this prediction: "in the next five years, the world's total car market will increase by about five million cars to about forty-two million cars. Almost two million of them will originate in the Asia-Pacific market, a third in China and two-thirds in the remaining emerging markets."[7]

Nonetheless, there are two caveats to these optimistic predictions prior to the Asian crises. First, there has been a fall in wholesale prices across Asia, with significant deflation in Japan and China—strong indicators of general overcapacity. Since the crisis, overcapacity has been especially evident in the auto market. Second, the Chinese market in particular is still largely a *potential* one. Demand is still poorly understood and currently small, with sales of about 1.5 million vehicles a year compared to U.S. sales of cars and light trucks of about sixteen million units a year. Since the 1997 crisis, savings rates have risen and spending has decreased. Analysts suggest that workers fear that they will lose their jobs as industrial restructuring accelerates. Incomes are still relatively low; auto financing does not exist; and distribution continues to be chaotic. Nonmarket barriers include a wall of tariffs and regulatory protection that will only come down when China enters the World Trade Organization (WTO). We discuss these factors in greater detail below, beginning with a broad examination of the Asian market in general and a more specific discussion of China.

Market Position

In analyzing the market forces at work in the Asian auto sector, we follow Michael Porter's "five forces model": (1) the bargaining power of suppliers; (2) the bargaining power of buyers; (3) the threat of substitutes; (4) the degree of competition; and (5) the degree of globalization in the industry.[8] We rank the relative importance of these forces for European firms in the Asian context and find the last two factors to be the most significant ones in shaping the success of market penetration.

Local suppliers have little bargaining power vis-à-vis the major auto firms, because Asia lacks a network of suppliers capable of bargaining with long established and technically savvy Western firms. In fact, the relative absence of local suppliers provided a barrier to entry into these markets. Because local content requirements were mandated throughout Asia, firms wishing to locate production facilities in the region had to provide years of technical training, certification processes and technology transfer. Nonetheless the relative absence of local suppliers provided a long-term bargaining advantage for western firms. As individual firms provided firm-specific training, certification, and technology to local suppliers, the latter's dependence on those firms was assured.

The growing demand for automobiles in these markets means comparatively weak bargaining power for consumers. In the mature North American and European markets, consumers can choose among hundreds of models, forcing manufacturers to move rapidly through product cycles, thus contributing to the increase in production costs. High demand gave local buyers in Asian markets little say over the level of product differentiation; Western firms were therefore in a good position to slow product cycles and reduce manufacturing costs.

As noted above, however, the level of actual demand in China should not be overstated. Most automakers that enter the Chinese market can count on a low volume of sales at the outset, with the expectation that demand will eventually increase. They therefore must be willing to produce a number of product types to find a wider initial market. GM Shanghai, for example, produces only thirty thousand Buicks a year but plans to add the production of a minivan and small car to achieve economies of scale.

As in other parts of Asia, auto manufacturers in China do not face a "threat of substitutes." Public transportation is underdeveloped, even in the cities. Four factors are responsible for reducing the threat of public transport as a substitute for automobiles: first, public transport is not efficient in serving areas with low population or employment densities. Low usage means infrequent service, and infrequent service, in turn, deters users. The kind of demographic fragmentation that characterizes most of Asia is an almost insurmountable challenge for public-transport systems.

Second, the rapid growth of Asian economies has changed travel patterns as new growth areas have sprung up. Fixed transport systems, such as rail lines, quickly become obsolete under conditions of rapid growth. The sunk costs that characterize mass transit systems are simply too high in areas where the shift from rural to urban demographics is rapid, and also too high when growth in developing economies is low.

Third, many types of public transport have high opportunity costs. Flashy rail systems can consume resources that could serve far more people if devoted to improving bus travel. And finally, the preference for public transit decreases as incomes rise; at the same time, consumers increasingly prefer autos over motorbikes.

The issue of competition in Porter's model has been complicated by transnational mergers and acquisitions. In the 1980s and 1990s, the auto industry was slow to respond to pressures for mergers despite excess capacity. Consolidation, automakers felt, would undermine brand recognition and loyalty, considered in the industry to be a key weapon in the fight for market share. And as the Japanese auto industry grew stronger, international competition among national firms intensified.

While the auto industry as a whole tried to fend off consolidation, Japanese firms came to dominate Asian markets, with significant and growing European market penetration, especially in China and Taiwan. Local Asian manufacturers also increased market share from the 1980s. In Malaysia, for example, Proton and Peruda, both Malaysian firms, increased their market share from 15 percent of all automobiles sold in 1987 to over 30 percent by 1996, with Japanese manufacturers still maintaining their hold over 60 percent of the sales market. The vehicle market in the Philippines was divided between Japanese and South Korean manufacturers; Japanese firms had an 80 percent market share, while South Korean firms controlled 15 percent. In the Indian market, Suzuki, through its joint venture with the state-owned Maruti holding company, had been able to increase its market share from 33 percent in 1987 to over 43 percent by 1996; the remainder was divided between European, Indian, and other Japanese manufacturers. More than 90 percent of the Indonesian market between 1991 and 1996 was controlled by Japanese manufacturers. Japanese and South Korean manufacturers each continue to control 95 percent of their domestic markets, although the import share in the Japanese market increased from 1 percent in 1980 to over 5 percent by 1994. The only market that U.S. firms have successfully penetrated is Taiwan, where Ford increased its market share from 19 percent in 1991 to over 23 percent in 1996. Nonetheless, Japanese manufacturers still control over 50 percent of the market.[9]

China was the only market that Japanese firms had not conquered. The Chinese regime had raised a number of barriers to entry for Japanese firms, and by 1985, Japan controlled only 20 percent of the market, with the remainder held by state-owned Chinese firms. Thus, both because of its potential for rapid growth and the small Japanese presence, American and European automakers have considered China to be the last market frontier.

Driven both by continued overcapacity in the 1990s and by intense competition in the Asian market, cross-national consolidation began to eclipse cross-national competition. Indeed, the problem of overcapacity had grown worse: in 1999, the average worldwide plant utilization was only 69 percent, compared with 80 percent in 1990. By the end of the decade, most national firms were pushed by lower profit margins to merge. A few examples illustrate the changed structure of the industry at the turn of the century: Ford held a controlling 33 percent interest in Mazda, and General Motors (GM) had acquired 49 percent of Isuzu and almost 10 percent of Suzuki Motors. In 1999, GM bought 20 percent of the Subaru car business of Fuji Heavy Industries. And DaimlerChrysler's purchase of a one-third interest in Mitsubishi Motors merged German, American, and Japanese firms into the third largest auto company in the world.

Now, the competition is not between national firms but between consolidated firms, often acquired for their competitiveness in specific market niches. It is too soon to tell whether these mammoth firms will simply be holding companies for the smaller manufacturers, but it is likely that the mergers will diversify production profiles so that firms can position themselves more competitively worldwide. DaimlerChrysler, for example, which holds a significant share of the world market for trucks and large cars, needed Mitsubishi's smaller models to fend off growing competitive assaults on its overall market share from Ford, Volkswagen, and Opel. And a controlling stake in Mitsubishi with its strong position in Asia would give DaimlerChrysler an immediate presence in Asian markets. Indeed, DaimlerChrysler's CEO, Jurgen Schrempp, made it clear that he wanted the surging Asian market to provide a quarter of sales by 2010, compared with 3.2 percent in 1999. Similarly, when General Motors acquired Subaru, it held only 1 percent of the Asian market, anticipating that Subaru would create a wedge that would permit GM more access there.

In addition to this cross-national consolidation, increased globalization of the automobile industry further changed the nature of competition. As Asian economies began to liberalize their economies, lowering tariff barriers and phasing out local content requirements, local producers could realize economies of scale by producing parts for a number of companies rather than just one or two. And auto firms could buy parts from the most efficient producers and ship them to their factories worldwide, rather than attempt to acquire all parts for each factory from small suppliers in each country. Indeed, increasing trade liberalization permitted Renault-Nissan and DaimlerChrysler to pursue strategies of volume production across at least two regional markets.

Even more radical strategies have been envisioned. The large manufacturers have begun to subcontract the design and production of entire subassemblies, such as brakes, steering, and suspension. For example, companies such as Valmet, a Finnish engineering company with roots in paper-making machinery, Magna, a Canadian parts company, and Steyr-Daimler-Puch in Austria are outsiders who have begun to make subassemblies for established companies like Porsche and DaimlerChrysler. Some analysts predict that the large consolidated companies might even begin to shed some of their assets to parts suppliers. The process of globalization is still in its infancy.

The apparent globalization trend in the auto industry may conflict with one of the key nonmarket strategies needed for success in the Chinese market: connections with local officials who have access to both political and economic resources that the auto industry may need. The more parts and sub-assembly suppliers in the region, the larger the political constituency supporting the foreign automaker. As we shall see below, the success of VW can in large part be attributed to its diverse local suppliers. Below we discuss this and other nonmarket factors and strategies that shape success in the penetration of China's market.

Nonmarket Position

In China, the strong arm of the authoritarian government and the continued use of five-year plans for industrial production and technology transfer in aid of the creation and growth of a domestic auto industry is the most important factor in any nonmarket strategy for market penetration. The Chinese government injected $2 billion over the 1995–2000 period into the auto industry through subsidies, preferential treatment in loans, import duties, overseas funds, and the like, to increase production consolidation and rationalization and to foster technological innovation.

> This state-led path of automobile production development, which increased from 222,000 units in 1980 to 510,000 units in 1990 to 1.45 million in 1995, has been explicitly directed under the auspices of an ambitious five-year plan which aims to change the present scattered structure of auto production, and to build large-scale auto enterprise groups with strong competitive powers, so as to realize the economy of scale . . . to change the backward, passive situation of the development of auto products; to build up independent, initiative R&D system . . . and to face the new circumstances of gradually connecting tracks of domestic markets with international markets.[10]

Having said this, however, the most crucial aspect of the nonmarket environment is *how* the resources are distributed in order to build the capability to meet the central plan targets. This entails a discussion of the institutional characteristics of the Chinese reform process in foreign direct investment (FDI), involving Sino-Foreign joint-ventures, center-local governmental relations, and informal administrative networks. Although conventional wisdom stresses the increased power of China's coastal regions over their central government during the period of economic liberalization, there is a lively scholarly debate about the consequences of this shift in power. Susan Shirk has argued that the internal Chinese Communist Party (CCP) power conflicts have allowed the regional party leaders to increase their power base and allocating mechanisms at the expense of the center,[11] thus providing an increasingly robust political base for the continuation of the reforms. Barry Naughton has argued that the logic of the reforms has been primarily economic—that is, based in the central state's need for an increased revenue base as well as the goal of accelerated national economic development.[12] Gabriela Montinola, Yingyi Qian, and Barry Weingast have argued that the increased devolution of power from the center to the regions has produced a market-preserving form of federalism, where the different regions have engaged in jurisdictional competition in order to generate market-friendly policies and outcomes.[13] Because of this, the relationship between foreign firms and local political officials is crucial to the firm's success. Steven Solnick and Gordon White, however, caution that although the center has seen its power reduced, it still maintains an overwhelming amount of control over high-profile investment projects.[14]

In this changing institutional context, informal *guanxi* networks within the center-region power relationship exercise an important influence on the success of Sino-foreign joint ventures.[15] Although the Chinese central state may have conceded some of its monitoring and enforcement mechanisms to its regional subordinate units, this loss of power has not been symmetrically distributed—*guanxi* networks have mattered more in some regions than others in terms of their ability to procure resources from the center toward facilitating joint ventures and accelerating the process of market-based economic development at the regional level.

Firm Position

Nonmarket Competencies.   The host state is instrumental in developing a coalitional base composed of local and regional governmental officials for automobile production.[16] As we shall see below, Volkswagen's successful management and negotiating strategies at each of these levels in setting up a joint venture in China were honed during the Cold War, when Germany established a number of joint ventures in Eastern Europe.[17] Studies of West European market penetration of Eastern Europe and the Soviet Union during that period suggest the importance of nonmarket strategies, particularly strategies in negotiating joint ventures.[18] George Halliday, for example, examined the Volga and Kama River joint automobile ventures with Western firms and demonstrated the importance of demanding high quality inputs in contract negotiations with Soviet officials.[19] The contractual stipulations for special plastics and metals and high-octane gasoline and service facilities drew inputs away from traditionally high-priority economic sectors, including the military sector. Success in this venture suggested that auto firms would also be successful if they made similar demands in negotiations with China. Because they have long experience in joint ventures with enterprises in centrally planned economies, European firms can use that experience in penetrating the Chinese market.

European automakers also face constraints and opportunities in their home environment which shape strategic decisions. The tripartite alliance between labor, business, and government is both a blessing and a curse. Preoccupied about unemployment, European governments and labor unions have traditionally resisted efforts on the part of auto makers to produce abroad; yet when this resistance is overcome, the tight relationship between banks and corporations in both France and Germany has provided the financial backing for international ventures. Indeed the fact that bankers and corporate CEOs sit on each others' boards of directors gives European firms a financial edge over their American counterparts.

In the last years of the 1990s, European auto firms have tended to focus their attention on the European Union and its regulatory environment. Indeed, European auto firms are tailoring their mergers and production bases in Asia to meet EU standards and compete within the boundaries of nonmarket constraints in Europe. For example, Daimler-Chrysler needed more fuel-efficient models in order to meet the European Union's directive to reduce fleet-average fuel consumption to sharply lower levels by 2008. Mitsubishi's range of ultra-efficient gasoline direct-injection (GDI) engines would permit DaimlerChrysler to conform to the new standard.

Market Competencies. European auto firms' competencies and market strategies left them at a distinct disadvantage relative to their Japanese competitors. By 1985, Japanese firms successfully overwhelmed and penetrated all other Asian markets, save South Korea and China. Expansion, particularly in North America, allowed them to establish substantial deep pockets—that is, growing cash reserves enabled Japanese firms to compete in Asian markets. With reference to cost and quality, their product sizes and low prices furthered their competitiveness and market share in Asia. In sum, Japanese firms' early market presence, competitively priced and durable products, and successful global expansion, in both production and distribution, created significant strongholds in Asian markets.

However, European firms had an important advantage in China. Much like the earlier movement of Japanese firms into Asian markets, they were counting on the timing factor. Both Peugeot and Volkswagen benefited from the first-to-market advantage, notwithstanding the Beijing Jeep project that was increasingly proving uncompetitive. Since Japanese manufacturers did not enter the Chinese market until the mid-1990s, European firms had fewer initial competitive pressures. Furthermore, constant intervention by the Chinese state meant that Japanese imports would face increased tariffs and duties. In effect, European firms exploited their original advantage from early entry to gain resources and build deep pockets, so as to generate a local, and possibly regional, stronghold. Finally, in terms of cost and quality, the European automobile firms were at less of a disadvantage than their American competitors. Their product lines and their production techniques also allowed them to compete with Japanese automobile manufacturers in a market where Japanese firms did not have an overwhelming advantage.

Firms faced market and nonmarket constraints and opportunities on the supranational level, domestically, and in China at both the national and regional levels. A firm could lobby its home state and multilateral bodies as well as implement a myriad of market strategies. But as Vinod Aggarwal argues in chapter one of this volume, what is truly important for firms is the fit between their core competencies and the institutional environment. Below we examine that fit in more detail in our comparative case study.[20]

III. VW and Peugeot in China

Background

For both Volkswagen and Peugeot, the decision to invest in production facilities abroad was driven by the evolution of consumer and labor markets in the auto sector at home. By the early 1980s, the European auto market was considered very mature with well-formed and rigid consumer preferences. Profit margins were low and growth opportunities were limited by demographics and existing income distributions. Industry officials were increasingly concerned with the growing ability of their Japanese competitors to continuously increase their market share on a global level. These considerations were exacerbated by increasing labor costs and slowly decreasing worker productivity characteristic of the European automobile sector in the late 1970s and early 1980s.

In particular, the long-standing bargaining power of the German trade unions placed a significant constraint on VW's ability to remain internationally competitive, especially in light of growing Japanese competition.[21] Indeed, the signing of the first VW deal came on the heels of the 1984 labor union strike that cost VW 160,000 units and led to production losses that decreased profits by 500 million Deutsche marks (DM) ($228.5 million).[22] Between the early 1980s and mid-1990s, hourly wages in the German auto industry (including taxes and social welfare costs) increased from $24.26/hr. to $39.39/hr.[23]

As noted above, part of the strategy to counter these losses was to step up the internationalization of production facilities through the acquisition of auto firms abroad and the construction of new plants in target markets. China was an important target. As Carl Hahn stated in a 1989 press conference at the 59th Geneva International Motor Show, "in light of deteriorating labor costs in Germany, we consider these Chinese ventures to be most important for us as far as the long-term future is concerned."[24]

An important key to a strategy for market penetration is knowledge about the preferences of officials in the host country who control market access. Chinese officials at the national level were most interested in the acquisition of auto technology for building a domestic auto industry. The creation of foreign production facilities would also contribute to the emergence and growth of components, supply, distribution, and services networks. Technology transfer would supply managerial skills, the influx of capital, the development of infrastructures, and access to export markets that could provide foreign exchange.[25] Officials also expected a multiplier effect: as the official Chinese news agency put it, "Sino-foreign automobile joint ventures are playing a vital role in upgrading the country's auto industry [because] they are helping the country bring up a new generation of automobile workers and management personnel, and forcing a large number of enterprises in related industries to reconstruct their production to meet the new needs."[26]

Two additional concerns dominated in the Chinese preference for contracts with European and American firms: the loss of foreign exchange with the rise of Japanese imports and increasing economic dependence on Japanese auto firms. Contracts with European firms and intended technology transfer would stimulate the construction of locally-based production facilities that would decrease the need to generate foreign exchange for auto imports.

Organizational Strategy and Tactics: Joint Venture Deals

Equity, investment, technology transfer, and production volumes led to the Chinese government's invitation to Peugeot and Volkswagen to bid for joint venture contracts. The first agreement was signed with Volkswagen on October 11, 1984. It was preceded by active lobbying by the German federal government, both at the national and at the Laender level.[27] Indeed, the last official visitor to China before the deal was signed was the Economics Minister of Lower Saxony, the home state of Volkswagen.[28] The equity investment in this deal was as follows: VW would invest 50 percent, the Shanghai Tractor and Automobile Corporation would cover 35 percent, and the Bank of China would provide 15 percent of the total. The venture would establish production facilities for the Santana model, with the goal of producing twenty thousand cars by 1988 and with the final goal of producing one hundred thousand cars by 1992. The agreement also called for the transfer of enough technology, training and equipment to raise the local component production to 90 percent. The agreement also called for the opportunity to export up to eighty thousand engines back to Germany by 1990.[29]

A similar agreement was signed with Peugeot five months later on March 15, 1985. Equity investment shares were divided as follows: Peugeot would invest $5.6 million; the Guangzhou Automobile Manufacturing firm would provide $11.6 million, the China International Trust and Investment Corporation would provide $5.1 million; the Banque Nationale de Paris $1 million, and the International Finance Corporation would provide equity participation of $2 million and a loan of $15 million. Peugeot agreed to provide management direction with equity investment of only 22 percent. The production of Peugeot's 504 pick-up model and of the 505 sedan would begin in 1988 with an initial goal of fifteen thousand units by 1988 and with a maximum targeted goal of thirty thousand units by 1993. Peugeot decided on the 504 and 505 models, because company officials predicted that the need for family and private cars in China would remain small for a number of years.[30] There was also an agreement to transfer 90 percent of Peugeot's technology used in the production of these models, including engines, stamped body parts, and axles.[31] This technology transfer was intended to cover Peugeot's 22 percent investment share in the form of licenses, equipment, engineering, and knowledge.[32]

China signed additional agreements with both firms in the early 1990s. Again, VW was first. VW agreed to produce the Golf sedan in the city of Changchun in the province of Jilin. Sixty percent of the joint venture, called the FAW-VW Automotive Company, would be owned by the Chinese government and 40 percent would be owned by VW. The initial investment would be DM600 million ($274 million), and the projected total investment would reach DM1.5 billion ($685 million), including the construction of a car assembly, engine, and gearbox plant. The venture was initially financed by a $420 million, eight-year loan, syndicated jointly by Commerzbank and the Hong Kong unit of the Bank of China.[33] The plants were designed to reach a full capacity of 150,000 units.[34] Indeed, statements by Chinese officials went as far as to argue that this increased cooperation in Sino-German auto production represented "the first step towards the target of stopping importation."[35]

Peugeot officials recognized that they had misjudged the market for passenger cars and made plans to compete directly with VW. In December of 1990, they began negotiations with the Chinese government to establish a Citroen plant in the city of Wuhan in the Hubei province. This venture would produce the Citroen ZX, a small-to-medium-sized car, designed to compete directly with the VW Golf. Citroen proposed to hold a 30 percent equity stake, with the long-established Second Automobile Works of China holding the rest of 4 billion yuan ($483 million) investment.[36] This deal was finalized in April 1992 with significant backing from French state-owned banks, with the French government directly providing 1.7 billion French francs (FFr) ($231 million) in low-interest loans and FFr 1.2 billion ($163 million) in buyer credits guaranteed by Coface, the French export credit agency, and FFr 1.1 billion ($150 million) provided by Chinese banks.[37]

Market Strategy and Tactics

Localizing Production: The Multiplier Effect.   Localizing production has both potential positive and negative benefits. We discussed the efficiency and standardization concerns above, and noted that consolidated auto firms are attempting to build regional (and even global) economies of scale through a rational division of labor among parts suppliers in the region. On the other hand, the localization of production and tight linkages between production networks and local suppliers can bring positive benefits. With local suppliers, a production base is created to launch an export strategy in the region—a more cost-efficient strategy for regional market penetration. Local production also gains favor with local authorities by stimulating the local economy, raising the profile of local officials, and permitting flexible production schedules. In short, business officials can more quickly respond to market demands when its suppliers are in close contact.

Localizing production, however, has its risks in terms of quality control. Indeed, firms must face trade-offs between efficiency, currying favor with political elites, and the final quality of the product. Quality products are the key to long-term maintenance of market share. How did VW compare with Peugeot in managing this trade-off?

VW initially localized production in China in order to escape the constraints set upon it by the state allocation of import licenses.[38] This motivation, of course, was consistent with the preferences of the Chinese government; the goal was to create a locally-based system of components and parts suppliers that would further reduce the costs of shipping the parts from Germany and thus make the Chinese operation increasingly more cost-efficient in order to begin exporting in the near Southeast Asian markets. Substantively, the Shanghai VW joint venture established a captive supplier policy, encouraging use of suppliers within its locality. Nearly all of its suppliers were under the auspices of the Shanghai Auto Industry Corporation, the joint venture's Chinese partner. In turn, the supplier qualification process took up to four years and required approval for quality control from VW's German headquarters.

In contrast, Peugeot established an open supplier policy precisely because of lack of established Chinese suppliers that met the quality standards of the French firm. Thus, instead of following the VW example and establishing and training local components suppliers, Peugeot opted for the import solution, continuing to rely heavily on suppliers in France.[39] What made the difference were the corporate strategic experience elsewhere and the corporate culture. VW had historically dedicated itself to a strategy of localizing content in all of its subsidiaries throughout Europe and Latin America; it was therefore able to draw on this experience in China. Peugeot, in contrast, had no such experience.[40] Out of the 120 suppliers that Peugeot wanted to locate in China, only thirty had effectively done so by the end of 1994.[41] In contrast, VW had one hundred components and licensing and know-how agreements and forty joint ventures since it began local production, and had another thirty joint ventures for parts and components under negotiation, effectively building the largest local supply network of any foreign manufacturer in China.[42]

Perhaps more importantly, the localization of production within China allowed VW the flexibility to produce at market levels. This allowed them to not only avoid having to submit to government-imposed production quotas, but also the need to generate the necessary foreign exchange for the importation of auto parts. In contrast, Peugeot maintained very low local content percentages in its production, even though it faced the same if not higher, import license quota constraints. This problematic nexus between content localization and import licenses was particularly acute for Peugeot. It aimed for the production of eight thousand units in 1990 but only acquired import licenses for forty-seven hundred units. It was not until October 1990 that the licenses for more components were granted.[43] While VW quickly achieved its local content goal of 90 percent, Peugeot never reached local content of more than 60 percent.

Distribution and Service Networks.   The creation of extensive distribution networks creates a powerful competitive advantage in that these networks act as a stronghold against market penetration by competitors. In China in particular, multiple distribution centers created before market liberalization could act as a barrier to potential market entrants when the market was opened. These networks, of course, also provide efficiency gains by permitting the product to reach the market quickly, thus assisting in maintaining targeted sales volumes. Extensive service networks pay off in reputation gains: they maintain product quality and longevity and thus increase brand loyalty.

VW officials were conscious of these benefits and created an extensive distribution network of over four hundred centers that stretched across the entire country.[44] Peugeot's distribution network was significantly smaller than VW's, which played a role in Peugeot's inability to generate a profitable sales volume. Indicative is a quote from an executive of a Western firm in China who was attempting to buy a Peugeot car: "When my company wanted to buy a Peugeot, there was no way to do so in Northern China. We had our driver sent to Guangzhou to get the car and drive it to Beijing."[45]

VW's extensive distribution network was accompanied by an equally extensive service network of over two hundred service stations throughout the country. These stations proved to be crucial in increasing the reliability of VW automobiles, given the rudimentary and infrequently serviced nature of the Chinese road system. VW has at least one service center in every region of the country, from Tibet to Shanghai; its rule of thumb was that if there are more than two hundred cars in one region, then there was to be at least one service center.[46] This reputation for paying attention to consumer needs was a first in the Chinese market. In contrast, Peugeot's service network of less than one hundred stations, combined with an unfortunate geographical selection of the plant's location, further increased the costs of repairs and monitoring.

Finally, VW's customer loyalty was bolstered by the introduction of new models to accompany the original VW Santana model. Peugeot, on the other hand, remained within the old framework of the 504 and the 505 models, even when there were signs that these models were not particularly in demand by the Chinese consumers as competing firms began to introduce newer and sleeker lines.

Human Capital.   Perhaps the most important market strategy was VW's emphasis upon human capital, both in management and in employment levels.[47] VW's management policy was to pair Chinese and German managers in such a way as to increase the learning curve of the Chinese managers. Furthermore, the German managers did not assume a patronizing tone toward the Chinese managers and the Chinese managers did not resent their better paid foreign counterparts. As we shall see below, Peugeot's strategy was quite different.

With regard to labor, VW faced the same problems that Peugeot faced: ill-disciplined workers who were grossly underproductive and often lacking the basic skills. The VW strategy was to create training institutes and workshops, both in China and in Germany. The training process was long, coordinated, and extensive: young workers were recruited from high school and given three-year courses, which included classroom lessons and practical training in such areas as machinery, welding, and forging. After this training, they were given permission to work on the assembly line. Some of these workers were then sent to a two-year program in management training in Germany and upon their return to China, they were placed as assistants to managers in order to gain some practical experience. If their performance was satisfactory, then they were promoted to managerial positions. In the words of Fang Hong, senior engineer and managing director at the Shanghai VW plant, VW's motto was to "train the people, organize the people, and motivate the people."[48] These training efforts were, of course, aimed at quickly increasing the technical know-how and the productivity of the Chinese auto workers.[49]

The creation of more training centers than were envisioned in the initial contract and the establishment of an independent R&D center spoke volumes about VW's commitment to worker training.[50] The R&D center would be supported by annually reinvesting 3 percent of the company's turnover. In the words of Martin Posth, the chairman of VW Asia-Pacific Group, this strategy would permit VW to "come up with cars engineered and designed between Germany and China with the latest technology, produced in China and exported to the Southeast Asian market, which Japan and South Korea now dominate."[51]

In contrast, Peugeot depended primarily on expatriate French managers and maintained limited interaction between Chinese and the French managers. The lack of contact between French and Chinese managers made it difficult for Peugeot to guarantee the productivity and the quality of the plant's workers. And without the "socialization" that comes with extensive training, Peugeot managers found it difficult to overcome problems in labor negotiations, particularly in its effort to link pay with performance. Its Chinese partners feared that the workers would not support any paying scheme that would increase intra-firm employee income inequalities.[52] In contrast, the VW's training program provided instruction on the link between pay and performance. As Fang Hong stated, "at Shanghai VW, we pay for the job and will only promote according to performance. If workers do well, they will get more wages, and will stand a better chance of being promoted."[53]

Nonmarket Strategy and Tactics

The importance of nonmarket strategies cannot be overemphasized in the Chinese context. The need to deal with the Asian "developmental state" is magnified in the Chinese case. Recall the discussion of center-periphery relations and the importance of firm relationships with local and regional officials in privileged regions, and the importance of understanding those officials' relationship with the center. The difference in region-level power greatly affected the survival and success of both joint ventures. Given a national context of politically determined preferential policies, it should come as no surprise that nonmarket strategies have been particularly powerful in determining the success or failure of business ventures.

The use of nonmarket strategies was particularly important in the Chinese automobile industry. Automobile production was considered by Chinese officials to be crucial to the growth and development of the national economy. Thus Western firms were compelled to deal directly with Chinese government officials, both at the central and at the regional-local level. The Chinese central state determined the production levels of the foreign auto firms, the number of foreign participants in the auto sector, the price level of the products (something that determined not only the profitability of the joint ventures but also affected the ventures' long-term economic viability), and the allocation of import licenses for sorely-needed components and spare parts that could not be efficiently and effectively produced in China. Indeed, during the early and mid-1980s, the central state set both automobile plant-level production targets as well as purchase orders for these production levels.[54] Furthermore, until 1994, the state provided the main market for autos produced in China.

For these reasons, both firms recognized the importance of political strategies aimed at government officials. However, while the German executives of VW were an almost constant presence in China, Jacques Calvet, the Peugeot CEO, did not visit China until 1997.[55] In contrast, Carl Hahn, the VW CEO, was in China on more than five occasions, each time stressing the importance of the Chinese market to the German firm as well as the need to create a fully localized product. Hahn made public statements to the effect that China represented a solid foundation as part of VW's long-term strategy and that VW had entered China with the aim of providing support not as a trader, but as a partner.[56]

The Presence of the State.   The German state, both at the local and at the federal level, was continuously and increasingly present in VW's Chinese venture, whereas the French state, especially the central government, remained conspicuously absent. The German government actively supported German firms in their efforts to penetrate the Chinese market. Indeed, the German government had discussed with Chinese officials the possibility of joint ventures for seven years before the 1984 agreement with VW was signed; it was the first Chinese joint venture with a European firm.[57]

High-ranking German officials made numerous trips with the explicit purpose of formally signing agreements and promoting even more joint projects. Indeed, Helmut Kohl accompanied Hahn to China for the signing of the Shanghai deal.[58] Close personal contacts paid off: for example, when the German parliament voted for the recognition of Tibetan independence, the Chinese government cancelled a visit by German Foreign Minister Klaus Kinkel, but did not alter the VW investment schedule. Kinkel was scheduled to visit China in July 1996, but after the Bundestag vote, Chinese officials postponed his visit until October, after what the

Chinese government claimed was the restoration of "normal, healthy relations with Germany." In the intervening period, according to the Chinese Foreign Ministry, the Germans had demonstrated that they were willing to discuss human rights issues "in a nonconfrontial [sic] manner, based on mutual respect and equality."[59]

Meanwhile, François Mitterand maintained an open policy of support toward Taiwan, which extended as far as allowing the sale of Mirage jet fighters (something that could alter the military balance between Taiwan and China) to the Taiwanese. Further, he openly criticized the human rights abuses of the Chinese regime. It is instructive in this sense to observe how Jacques Chirac sought to change this policy perception by pushing for closer relations with China.

Technology Transfer and Quotas.   In the context of these foreign policy issues, firm relationships with government officials were also crucial. The Chinese had been adamant not only in their demand for total technology transfer at the end of the joint venture but also for the introduction of technologically advanced production and assembly lines. Again, VW came out ahead. Indeed, its use of extensive worker training, the voluntary creation of an R&D center, and the increased localization of production afforded it a much more respectable position in the eyes of the Chinese decisionmakers than Peugeot. Similarly, in terms of quotas (both production and import license ones) what mattered was the firm's willingness to demonstrate its goodwill vis-à-vis the Chinese authorities' plans and schedules.

Firm Relationships with Local Authorities.   Crucial in both dimensions of local nonmarket strategies was the role that the Chinese provincial and municipal authorities played in the determination of Beijing's policies toward the two firms.[60] Put succinctly, VW's operation in Shanghai was more favorably situated for the reception of preferential state policies than Peugeot's project in Guangzhou due to the increased power and access that the Shanghai party officials enjoyed at the highest central levels. As Burkhard Welkener, the deputy managing director of the Shanghai joint venture stated in 1989, referring to the Chinese local partners, "we have strong local support from the municipal government, which has invested over $210 million thanks to the taxes, from the vehicles sold."[61]

Given the importance of *guanxi* networks in the determination of economic policy in contemporary China, it comes as no surprise that the two most important political actors in China in terms of economic policy in the 1988–1998 era were Shanghainese mayors who were instrumental in spearheading the regional economic liberalization drive. Indeed, the increased representation of Shanghai-based politicians within the higher echelons of the central state apparatus as well as their rapid rise has led some analysts to speak of a "Shanghai Mafia."[62] Other analysts have stressed that the rise of Jiang Zemin and of Zhu Rongji in 1990 allowed Shanghai much more policy autonomy, which in turn led to Shanghai's rapid rise as a foreign direct investment recipient area.[63] This emphasis on picking the local partner of a joint venture on the basis of their political connections cannot be underestimated. As Martin Posth argues, "if you have an investment with a bigger risk, you will need someone to deal with all those politicians and someone who knows the environment. As an example, what are you going to do if you don't get enough energy? To whom do you speak? The local manager? . . . My partner has sorted out many problems."[64]

In the earlier stages of the project, the Beijing government allowed Shanghai a variety of flexible economic measures designed to provide the region with an inviting foreign direct investment environment. Most of these measures focused on foreign currency regulation, including the increase in foreign exchange allocated to joint ventures and the granting of an authorization to the city government to demand payment for part of the joint venture products in foreign currency within China, which provided the VW joint venture with a more flexible way of raising foreign exchange for import licenses in the earlier stages of the project.[65] Indicative of this support for foreign direct investment in the Shanghai region was the establishment of an International Businessmen's Advisory Council to provide input to local policymakers.[66]

Both Jiang Zemin and Zhu Rongji did their utmost to increase the preferential treatment that the Shanghai-based auto projects received from Beijing in terms of subsidies and financial assistance in moments of economic crisis. In contrast, the Guangzhou province found itself increasingly isolated in the decisionmaking circles of Beijing, a fact that played an important part in Peugeot's decision to place the Citroen factory in Wuhan province, where the project received strong central government support from the beginning.[67]

This crucial fact was aptly illustrated both in the 1989 economic crisis and in the 1993 central designation of the auto sector as a primary "pillar" (the most important and hence most heavily subsidized). In 1989, two issues were crucial: the aftermath of the Tiananmen Square demonstrations and an economic downturn that resulted in increased taxes on automobiles and a subsequent slump in the Chinese government's demand for them. Notably, this was in addition to the central government's $6,750 tax in 1988 on each car produced and a $5,400 Special Consumption Tax created in March of 1989. Bruno Gandeler, the general manager of the Guangzhou Peugeot joint venture, lamented then that, "apart from the problem of getting a purchasing right ticket, the customer has to pay yuan 200,000 ($54,000 [at 1989 rates]) for Peugeot 505 station wagon. We told them that they would kill the industry, but they ignored us. They say they are now considering canceling the new taxes, but everything in China takes a very long time, and if they do not hurry up, we are dead."[68]

The central government responded to this and other warnings about the fragility of the auto industry with an "emergency purchasing plan." The plan was formulated by the China National Automotive Industry, under the auspices of the Ministry of Materials and Equipment, with an aim to "help Sino-foreign joint venture automobile producers out of present difficulties."[69] As a result, VW received a bailout in 1989 from the Bank of China in the form of a 100 million yuan ($27 million, at 1989 rates) loan for expansion. At the same time, however, Peugeot was still negotiating with the Bank of China for a smaller loan after one of its partners, China International Trust and Investment Corporation (CITIC), had failed to increase its equity stake in the company. Additionally, VW was permitted both a higher production quota and a higher quota of import quota licenses.[70]

Even in the process of the "emergency purchasing plan" there were different purchasing quotas: the State Purchasing Commission bought fifteen hundred Santanas from the VW joint venture, five hundred Jeeps from the Beijing Jeep Corporation (the joint venture with AMC), and eight hundred Peugeots.[71] Both preferential actions were ascribed to the Shanghai Automobile and Tractor Industrial Corporation, Volkswagen's partner in the joint venture. Shanghai Tractor was one of the most influential municipal-level organizations in China, with direct links to the central government in Beijing, and it was assumed to have been the initiator of the "emergency purchasing plan."[72]

In 1993, VW was designated as one of the privileged firms in the auto sector pillar of the economy, meaning that it would receive increased state assistance and preferential treatment in the process of market liberalization. That meant that market barriers would be maintained as long as VW products could not compete successfully with other imports and other domestically-produced autos. Peugeot, on the other hand, was deemed a mere domestic producer, possibly as a result of the distrust for the provincial government's allegiances at Beijing.[73] Other preferential treatments included tax exemptions, policy-oriented loans, and priority in using foreign funds and listing in stock and bond markets.[74] All these policies allowed VW to increase its chances to achieve economies of scale production capabilities, which for some analysts proved to be a powerful tool in solidifying its status as the preeminent Sino-foreign automobile joint venture.[75] In short,

Figure 6.1   Chinese Automotive Production, 1959–1996

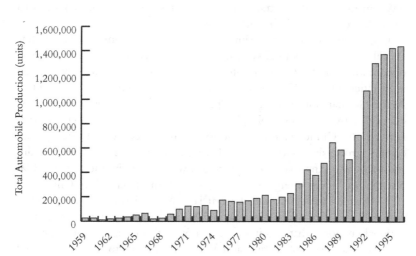

Source: World Motor Vehicle Data Program

VW greatly outmaneuvered Peugeot in that it was able to side with the most important and powerful provincial and local authorities in China.[76]

Finally, the VW management board publicly stated that China's entrance into the WTO should be delayed if the international community would not establish safeguards for the protection of domestic infant industries, like the auto industry.[77]

IV. Conclusion

For Peugeot, the China experiment ended in disaster and in the sale of its production facilities. Part of the failure was market driven: Peugeot cars were too big and uneconomical for the size and the needs of the Chinese consumer. Vertical production cycles were not implemented and there was an overdependence on French expatriate managers. Further, the distribution network was too small to increase the chances for product penetration. The Peugeot strategy had always been more oriented toward the Eastern European, Latin American, and South Asian (Indian) markets. Part of Peugeot's failure can be directly accounted for by its faulty nonmarket strategies: cooperation with the local authorities was slow to emerge, and VW's prior successful entrance in the Chinese auto market appeared to drive Peugeot out of the market.

Conversely, the VW experiment showcased the combination of successful market and nonmarket strategies. The incentives for the internationalization of the German auto firms' production facilities were used most efficiently by VW that has consistently generated over 70 percent of all the German auto firms foreign production.[78] There was a strong production network of assembly plants as well as components factories that fulfilled the Chinese requirement for domestic content. There was an equally strong distribution network that further increased brand loyalty and awareness. And that brand awareness was assisted by the fact that VW was the first major foreign auto firm to establish operations in China.

Nonmarket strategies were also crucial. VW officials developed close ties with Chinese state officials in a favorable foreign policy environment. In a national economy that requires at least a 50 percent stake in joint ventures and significant technology diffusion, this factor appears to have only helped VW. Given the different level of state support that these two firms were receiving from their national governments and given VW's prior market entrance and capture of significant parts of the markets, it should not come as a surprise that VW has come off better. At this moment, VW and its local joint ventures in China have moved beyond the production of the Santana in the South and have begun to include the production of more upscale Audis in the northern part of China. VW consistently maintains in excess of 50 percent of the local production and sales markets. This growth was maintained in the mid-to-late 1990s even in the face of the entrance of other auto firms in the Chinese market—indeed, for the first five months of 1997, VW saw its China sales increase by 22.5 percent over the previous year.[79]

It would appear that a discussion of business organization sheds light on another dimension of VW's success.[80] While Peugeot simply constructed assembly operations that made autos locally, it used only designs, processes, components, and management approaches developed in France, VW established plant complexes in strategic locations that fabricated vehicles using local management talent and local components suppliers. Although quality control dictated standard operating procedures for management and labor developed in Germany, VW made extensive use of local management talent and local suppliers. This type of organization benefited both market and nonmarket strategies.

VW's victory has not gone unnoticed by American and Japanese competitors currently entering the market. In Asia, U.S. firms are facing Japanese firms that have built an extensive set of cross-national production net-works—especially in Indonesia, Malaysia, and Thailand—that allow for both the effective penetration of those markets and for the hedging of fluctuations in the economic conditions of China while maximizing the benefits of free-trade zones and low wages. This is a position that both Japanese and U.S. auto firms face vis-à-vis VW in China.

Notes

1. For the most widely accepted definitions of the relevant aspects of the Asian "developmental state" see Wade (1994), and Johnson (1985).
2. See chapter 3 by Ravenhill in this volume for further information.
3. Karmokolias (1994), p. 4.
4. Karmokolias (1994), p. 7.
5. Maxton and Wormland (1995), p. 155.
6. *Financial Times,* June 25, 1997.
7. *Financial Express,* June 23, 1997.
8. Porter, M. E. (1980).
9. For more country-specific statistics observe the attached figure at the end of the chapter. It is based on data collected for the *World Motor Vehicle Data* (various issues). What is important for our analysis is the fact that the Japanese firms have remained the dominant firms in this regional market.
10. China Auto International Information (1997).
11. Shirk (1993).
12. Naughton (1995).
13. Montinola, Qian, and Weingast (1995).
14. Solnick (1995) and White (1993).
15. *Guanxi* networks have been defined as both vertical and horizontal networks that operate both in terms of symmetrical and asymmetrical reciprocity. For more on *guanxi* networks see Walder (1986) and Gold (1986).
16. Doner (1991).
17. See Harwit (1995).
18. See Crawford (1993).
19. See Halliday (1979).
20. See chapter one by Aggarwal in this volume.
21. For more on the bargaining power of German trade unions see Schonfeld (1965).
22. *Handesblatt,* July 13, 1984. All currency conversions in this chapter are of March 27, 2001, unless otherwise noted.
23. The increase of nearly 50 percent was significantly higher than all their competitors, with the possible exception of the Japanese. See *Automotive Industries* (August 1996), p. 45.

24. Reuters, March 10, 1989.
25. Osland and Cavusgil (1996).
26. Xinhua News Agency, June 28, 1987.
27. A *laender* is a state in the German Federation.
28. *Handesblatt,* February 28, 1984.
29. *New York Times,* October 11, 1984; *Financial Times,* October 11, 1984.
30. *Financial Times,* September 26, 1986.
31. *South China Morning Post,* October 18, 1985; *Financial Times,* September 22, 1986.
32. *Financial Times,* March 14, 1985.
33. *New York Times,* November 10, 1992.
34. *Financial Times,* November 30, 1990.
35. Reuters, November 20, 1990.
36. *Financial Times,* December 21, 1990.
37. *Financial Times,* April 10, 1992.
38. *Chicago Tribune,* January 14, 1994; *South China Morning Post,* November 20, 1995.
39. *China Business Review* (March 1994).
40. *Far Eastern Economic Review,* March 26, 1992.
41. *L'Usine Nouvelle,* April 4, 1994.
42. *Financial Times,* December 2, 1994; May 2, 1995.
43. *Les Echos,* August 1, 1990; *South China Morning Post,* October 19, 1990.
44. *South China Morning Post,* November 20, 1995.
45. *China Economic Review* (January 1997).
46. *Financial Times,* May 2, 1995; *IPC Industry Week,* July 15, 1995.
47. By late October 1985, a total of thirty-six VW training staff were already on the ground in Shanghai, headed by Hans-Joachim Paul whose previous job was as the division head of the VW factory in Kessel. See *Financial Times,* October 29, 1985.
48. *China Business Review* (September 1992).
49. Xinhua News Agency, June 28, 1987.
50. *South China Morning Post,* July 2, 1995.
51. *South China Morning Post,* November 20, 1995.
52. *Asian Wall Street Journal,* January 27, 1987.
53. *China Business Review* (September 1992).
54. Zongkun (1987), p. 234.
55. *Daily Telegraph,* March 12, 1997.
56. Xinhua News Agency, November 8, 1989.
57. Reuters, February 28, 1984.
58. *New York Times,* October 11, 1984.
59. *Financial Times,* October 23, 1996.
60. Reuters, June 21, 1986; Xinhua News Agency, January 18, 1987.
61. *Journal of Commerce,* April 10, 1990.

62. Miller (1996).
63. Chow and Fung (1997), p. 254.
64. *IPC Industry Week,* October 27, 1997.
65. Xinhua News Agency, July 15, 1986.
66. *Economist,* March 17, 1990.
67. *China Economic Review* (January 1997).
68. *Financial Times,* October 19, 1989.
69. Xinhua News Agency, October 29, 1989.
70. *South China Morning Post,* December 7, 1989.
71. *New York Times,* November 20, 1989.
72. *South China Morning Post,* December 7, 1989.
73. *USA Today,* September 10, 1996.
74. *China Daily,* July 12, 1995.
75. De Bruijn, Jia, and Konterman (1997), p. 357.
76. In this scenario, power refers to influence in policymaking.
77. *South China Morning Post,* July 15, 1997; *Daily Telegraph,* March 1, 1997.
78. *Automotive Industries* (August 1996), p. 45.
79. *Hong Kong Standard,* June 19, 1997.
80. Hixon and Kimball (1990).

References

Chow, Clement and Michael Fung (1997). "Profitability and Technical Efficiency in Shanghai." In *Advances in Chinese Industrial Studies,* edited by Sally Stewart and Anne Carver. Greenwich, CT: JAI Press.

Cusumano, Michael (1985). *The Japanese Automobile Industry: Technology and Management at Nissan and Toyota.* Cambridge, MA: Harvard University Press.

Cusumano, Michael A. and Kentaro Nobeoka (1992). "Strategy, Structure and Performance in Product Development: Observations from the Auto Industry." *Research Policy* 21 (3), pp. 265–293.

Crawford, Beverly (1993). *Economic Vulnerability in International Relations.* New York: Columbia University Press.

De Bruijn, Erik, Xianfeng Jia and Ina Konterman (1997). "Consequences of Change for Joint Ventures in China." In *Advances in Chinese Industrial Studies,* edited by Sally Stewart and Anne Carver. Greenwich, CT: JAI Press.

Doner, Richard (1991). *Driving a Bargain: Automobile Industrialization and Japanese Firms in Southeast Asia.* Berkeley: University of California Press.

Gold, Thomas B. (1986). *State and Society in the Taiwan Miracle.* Armonk: M. E. Sharpe.

Halliday, George (1979). "The Role of Western Technology in the Soviet Economy" in *Issues in East-West Commercial Relations,* U.S. Congress Joint Economic Committee. Washington, D.C.: U.S. Government Printing Office.

Harwit, Eric (1995). *China's Automobile Industry: Policies, Problems and Prospects.* New York: M. E. Sharpe.

Hixon, Todd and Ranch Kimball (1990). "How Foreign-Owned Businesses Can Contribute to U.S. Competitiveness." *Harvard Business Review,* January-February, pp. 56–5.

Johnson, Chalmers A. (1995). *Japan: Who Governs? The Rise of the Developmental State.* New York: Norton.

Karmokolias, Yiannis (1994). *Radical Reform in the Automotive Industry: Policies in Emerging Markets.* Washington, D. C.: World Bank.

Kenney, Martin and Richard Florida (1993). *Beyond Mass Production: The Japanese System and its Transfer to the U.S.* New York: Oxford University Press.

Maxton, Graeme and John Wormald (1995). *Driving over a Cliff? Business Lessons from the World's Car Industry.* London: Economist Intelligence Unit.

Miller, Lyman (1996). "Overlapping Transitions in China's Leadership." *SAIS Review* 16 (2), pp. 21–42.

Montinola, Gabriela, Yingyi Qian, and Barry Weingast (1995). "Federalism, Chinese Style: The Political Basis for Economic Success." *World Politics* 48 (1), pp. 50–81.

Naughton, Barry (1995). *Growing out of the Plan: Chinese Economic Reform, 1978–1993.* Cambridge: Cambridge University Press.

Osland, Gregory and S. Tamer Cavusgil (1996). "Performance Issues in U.S.-China Joint Ventures." *California Management Review* 38 (2), pp. 106–130.

Porter, M. E. (1980). *Competitive Strategy.* New York: Free Press.

Shirk, Susan (1993). *The Political Logic of Reform in China.* Berkeley, CA: University of California Press.

Shonfield, Andrew (1965). *Modern Capitalism: The Changing Balance of Public and Private Power.* New York: Oxford University Press.

Solnick, Steven (1996). "The Breakdown of Hierarchies in the Soviet Union and China: A Neoinstitutional Perspective." *World Politics* 42 (2), pp. 209–238.

Stewart, Sally and Anne Carver, eds. (1996). *Advances in Chinese Industrial Studies.* Greenwich, CT: JAI Press.

Tate, John Jay (1995). *Driving Production Innovation Home: Guardian State Capitalism and the Competitiveness of the Japanese Automobile Industry.* Berkeley, CA: Berkeley Roundtable on International Economy.

Tidrick, Gene and Chen Jiyuan, eds. (1987). *China's Industrial Reform.* Washington, D. C.: World Bank Publications.

Wade, Robert (1990). *Governing the Market: Economic Theory and the Role of Government in East Asian Industrialization.* Princeton, NJ: Princeton University Press.

Walder, Andrew G. (1986). *Communist Neo-Traditionalism: Work and Authority in Chinese Industry.* Berkeley, CA: University of California Press.

White, Gordon (1993). *Riding the Tiger: The Politics of Economic Reform in Post-Mao China.* Palo Alto, CA: Stanford University Press.

Zongkun, Tang (1987). "Supply and Marketing" in *China's Industrial Reform,* edited by Gene Tidrick and Chen Jiyuan, pp. 210–236.

Chapter 7

David and Goliath:
Airbus vs. Boeing in Asia

William Love and Wayne Sandholtz

I. Introduction

In an era when almost any kind of transnational business activity is taken as evidence of "globalization," for many companies little has changed. As they have for at least a century, international firms manage overseas production facilities and seek to expand their sales in foreign markets. In this sense, the difference between a globalized market and a collection of national ones is not at all clear. In the commercial aircraft industry, however, the term "global" still applies. Though production is still overwhelmingly national (Boeing, a U.S. company) or regional (Airbus, a European consortium), the only relevant market is international. No national market, and no regional market, is large enough to sustain a profitable aircraft industry, given the immense development costs that go into each model. Even the world market for aircraft is currently large enough to support only two manufacturers, Airbus and Boeing.

That said, different regions play different roles in the strategies of the aircraft producers. This chapter focuses on Airbus's strategies in Asia. Yet in order to understand the approach Airbus takes in Asia, we must grasp Asia's place in the broader international strategic context. From the perspective of Airbus, the major regions contribute in distinctive ways to an international strategy. Europe plays the part of a stronghold, a home market where Airbus can count on significant orders. For instance, when Airbus faced the challenge of breaking into the airliner market in the early 1970s with its first model, the A300, orders were few and small. In those lean years, the consortium could at least count on orders from the national flag carriers of the participating countries. The two major members of the Airbus group were France's Aérospatiale, and Germany's Deutsche Airbus. Both were receiving government subsidies for development of the airplane. Not surprisingly, the first orders for the A300 came from Air France and Lufthansa—the latter after some arm-twisting by the German Ministry of Transport.[1] The North American market is the world's largest and richest, and is therefore crucial to the success of Airbus. To reach the volume of sales necessary, the American market is indispensable. The Asian market, however, has been the fastest growing and will likely continue to show the most dramatic expansion well into the twenty-first century, despite the recent crisis.[2]

Asian markets play a crucial role in the strategies of both Airbus and Boeing. The growth of airline fleets in Asia will translate into substantial orders for new aircraft. Airliner purchases in Europe and North America will experience more modest growth. In these regions, air travel is already highly developed; new aircraft orders will often serve not to meet rising demand but simply to replace aging planes. Furthermore, Asian markets are attractive because there are no indigenous producers to compete with. To the extent that a "buy American" inclination exists in the United States, and a "buy European" sentiment in Europe, the fast-growing Asian markets offer a more nearly neutral arena.

The first section of the paper discusses the positional analysis, focusing on key features of the aircraft industry and markets. Technical and financial aspects of aircraft manufacturing, coupled with the nature of the international market for jetliners, define the broad structural constraints within which Airbus and Boeing must devise their strategies. Asian markets offer certain specific challenges, which the first section will describe along with a brief history of Airbus in the region. The second section of the paper addresses the strategic and tactical analysis, including assessments of Airbus's organizational, market, and nonmarket activities. The final section relates our findings to the book's general framework of integrated strategies in highly competitive industries, and offers some conclusions.

II. Positional Analysis

Airbus's strategies in Asia respond to technical and financial aspects of aircraft manufacturing, general features of the world market, specific traits of Asian markets, and distinctive characteristics and objectives of the consortium.

Market Position and Firm Position

The Commercial Aircraft Industry and the World Market.   According to the title of one book on the subject, building commercial aircraft is a "sporty game," a business only for people who like to gamble large sums at low odds.[3] It is an industry in which the technical and market uncertainties are as immense as the investments required to "play." Two firms compete in a market that is truly global. In this contest, Boeing is the established player, and Airbus is the successful upstart. The Airbus consortium came into being in 1970; the first Airbus plane entered commercial service in 1974. Figure 7.1 shows the dramatic fashion in which Airbus has closed the gap with its more established rival since 1974.

Of first importance is the technology-intensive character of aircraft development and production. As of 1985, research and development (R&D) expenditures in the aircraft industry equaled 17.5 percent of net sales value; only the electronics industry showed a higher ratio.[4] Though the civilian aircraft industry is not necessarily a major source of technological innovation, as a major consumer of high-tech parts and subsystems from a variety of industries, it generates "demand-pull" pressures for innovations across a broad front of technologies. Aircraft incorporate advanced electronics, navigation, communications, hydraulics, materials, and propulsion technologies. One of the most demanding and unpredictable challenges of modern aircraft manufacture, and possibly the crucial technical feature of the industry, is the integration of these diverse subsystems into an efficiently functioning whole.[5] Aircraft producers thus invest heavily in design, modeling, systems integration, and testing.

Even as they focus on the complex technical tasks of designing and producing a new aircraft, producers must simultaneously pay close attention to their customers, the various airline companies around the world. Not only must the airplane be technologically up-to-date and solve thousands of technical problems, it must meet the needs of the airlines in terms of passenger capacity, range, fuel efficiency, noise, ease of training and service, and so on. The aircraft makers thus consult intensively not just with a few

Figure 7.1   Airbus and Boeing Annual Deliveries, 1974–1999

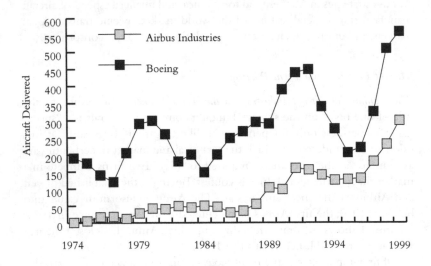

Source: For 1974–1998, Airclaims CASE Database; for 1999, Airbus Industrie (www.airbus.com) and Boeing (www.boeing.com).

likely customers, but with a broad range of airlines. They must try to blend the diverse airline requirements into a single design that will maximize the sales potential of the airplane. The pressure to "get it right" is immense, since each new airplane will have an extraordinarily long lifetime of production and sales. Boeing manufactured its 727 for nearly twenty years and has produced the 747 for over thirty; the Airbus A300 has been in production for over twenty-five years.

During its long lifespan, an aircraft continues to evolve through the incorporation of new electronics or engines, or through developing "derivatives." Derivatives take an existing model and either stretch or shorten it by adding or removing fuselage sections. Derivatives allow the maker to expand its aircraft family, by adding airplanes of somewhat greater or lesser capacity (in passengers or cargo), without designing new models from scratch. The basic design remains essentially the same (cockpit, controls, fuselage cross-section), though stretching often means upgrading the wings and usually requires new engines. The marginal design costs associated

Table 7.1 Development Costs of Selected Aircraft

Aircraft	Year Entered Service	Development Costs (millions of 1991 U.S. dollars)	Development Costs per Seat (millions of 1991 U.S. Dollars)
DC-3	1936	3	0.1
DC-6	1947	90	1.7
DC-8	1959	600	3.75
747	1970	3,300	7.3
777 (estimated)	1995	4,300	14.0

Source: Hayward (1994), p. 7.

with stretching are, of course, far less than those associated with designing an entirely new model. Derivatives thus allow the producer to amortize the heavy up-front costs over a longer period and in a new market niche. Economists David Mowery and Nathan Rosenberg estimate that the costs of stretching "rarely exceed 25 percent of the original development costs."[6] Even so, developing derivatives can be extremely expensive. For instance, Airbus is currently working on two derivatives of the A340, its four-engine, long-range widebody. The A340–500 will carry 313 passengers for up to 15,750 kilometers (about the same capacity but longer range than the basic A340); the A340–600 will accommodate 380 passengers over 13,900 kilometers. The development costs for the two new derivatives are reported to be about $2.5 billion.[7]

Because of the extreme demands of up-front design and testing, the development phase for any new model is long and costly. The design and development phase typically runs for four to six years. The costs incurred during this phase are sizeable, as shown in table 7.1. The $1.2 billion in development costs for the 747 represented more than three times the total market capitalization of Boeing at the time. Douglas spent a similar multiple of its total capitalization to develop the DC-10.[8] Airbus is currently in the middle of design and development work for its planned super-jumbo, the A380, which would carry a minimum of 480 passengers. The consortium estimates that development costs for the A380 will total $10–12 billion.[9]

Given the massive initial investment in an airplane, even after the successful flight of the first prototype, profitability is years away. Many—perhaps most—models never reach it. Laura D'Andrea Tyson cites an estimate by the Office of Technology Assessment to the effect that a new

Table 7.2 Total Orders for Selected Aircraft as of January 31, 1999

Airbus		Boeing	
Aircraft Type	*Number of Orders*	*Aircraft Type*	*Number of Orders*
A300/310	781	727	1,831
A319/320/321	1,923	737	4,250
A330/340	507	747	1,287
		757	966
		767	864
		777	419

Sources: Airbus Industry, "Our Market," at http://www.airbus.com; Boeing, "Commercial Airplanes Order and Delivery Summary," at http://www.boeing.com/commercial/orders/index.html

model produces "negative cash flow until about the seventieth unit," and does not break even until it reaches about six hundred in sales.[10] Airbus officials estimated in the early 1970s that the consortium's first plane, the A-300, would break even at 360 aircraft sold; by 1982 they had revised that estimate to between 850 and nine hundred.[11] Table 7.2 reports the total number of aircraft ordered for the various Airbus and Boeing models.

Over the lifetime of a specific model, the total number sold has an overwhelming impact on the production cost per unit. The more units sold, the cheaper it is to build the next one. A 1985 study by the National Academy of Engineering and the National Research Council expressed the relationship in the following way. Assuming total projected production of seven hundred airplanes, a 50 percent shortfall in sales (to 350 planes) would increase per unit costs by 35 percent.[12] Another way of putting it is to note, as Tyson does, that "the commercial aircraft industry is distinguished by unparalleled increasing returns to scale."[13] Among industry analysts, the consensual estimate is that every doubling of output decreases production costs by 20 percent.[14] To take an Airbus example, the first A300 required 340,000 worker-hours to produce the fuselage sections. For the seventy-fifth aircraft, the number had fallen to eighty-five thousand worker-hours and it would continue to drop to forty-five thousand.[15] Naturally, the learning effects acquired in production transfer from the original model to its derivatives (stretched and shortened versions). Aircraft manufacturing thus displays three important economies:[16]

- Static economies of scale (design, development, and tooling)
- Dynamic economies of scale (learning by doing)
- Economies of scope (learning transferred to derivative models)

The dramatically increasing returns to scale make pricing strategies extremely tricky and flexible. Though it is virtually impossible to determine the actual price paid by an airline for a specific plane, there are some plausible generalizations. One is that the price of an airplane bears no direct relation with the cost of making it. Manufacturers price their planes around the estimated average cost, assuming an extended production life in the range of four to six hundred machines.[17] Thus, aircraft producers clearly sell a new model at prices that are well below their actual costs, hoping to make up the difference as learning effects kick in and drive down marginal costs.

Still, for any given airplane, the odds are against the producer making a profit. The competition among aircraft makers can be devastating. For example, Lockheed and Douglas undertook a debilitating head-to-head competition in wide-body airplanes with three engines, the L-1011 and the DC-10 (later to become the MD-11). Both planes were launched shortly after Boeing introduced its 747. The world market simply could not support both the L-1011 and the DC-10, and the effort severely weakened both companies.[18] Ultimately neither company survived. Lockheed exited the commercial aircraft business in 1981, and Boeing bought McDonnell Douglas in 1997, leaving Boeing as the only American firm in the industry. In fact, the story in Europe's postwar aircraft industry has also been one of dramatic consolidation. By 1977 the last two British airframe makers (Hawker Siddeley and British Aircraft Corporation) merged to become British Aerospace (BAe). Aérospatiale was the product of mergers in France that left it the sole commercial producer by 1970. In Germany, five aerospace firms banded together in a consortium named Deutsche Airbus, which would handle Germany's share of Airbus work. The largest of the five, Messerschmitt-Beolkow-Blohm (MBB), would later take over one hundred percent of Deutsche Airbus. In 1989, the Daimler industrial group acquired MBB. Since Daimler's merger with Chrysler, the company is known as DaimlerChrysler Aerospace (DASA).

The Asian Markets.   Asia plays a distinctive role in the world aircraft market, and thus in Airbus's strategies as well. Whereas Europe constitutes a home market for Airbus (providing an advantage with some flag carriers), and North America offers the largest market, Asia is where the greatest growth will occur over the long run. In addition, Airbus officials have always suspected that breaking into the U.S. market would be slow and difficult, and that Asia would offer a more nearly level playing field. If anything, Asian airlines would prefer to see competition on the supply side of the aircraft market.

Until the onset of the recent economic crisis in Asia, the rapid expansion of air transport in the region offset stagnant or declining aircraft orders from North America and Europe. Growth in passenger traffic was slow in North America and Europe. In large part aircraft orders in those regions were replacing aging jets. Asia would need aircraft both for replacement and expansion. Thus early Airbus strategists thought in terms of establishing a major presence in Asia and connecting it westward to Europe along the "silk route." The first order from Asia (and the first order from outside Europe) came in September 1974 from Korean Air. The final link in the silk route strategy—the Middle East—fell into place in 1980 with forty-one orders from three airlines (Kuwait Airways, Middle East Airlines, and Saudi Arabian Airlines).[19]

The prospect of expansion in Asia still holds: Asia will be the fastest growing market for aircraft well into the twenty-first century. The crisis in Asian economies has postponed, but not negated, the growth. By 1994, the Asia-Pacific region accounted for 34.5 percent of scheduled international air traffic. That figure, according to the International Air Transport Association (IATA), is projected to rise to more than 50 percent by 2010.[20] Air passenger traffic in the Asia-Pacific region almost doubled in volume between 1994 and 2000, and is projected to double again by 2010 to almost 400 million. Whereas passenger traffic in the rest of the world is projected to grow by an average of 5.9 percent per year through 2010, in Asia it will expand by about 8.5 percent annually.[21] Airbus, in its 1997 publication *Global Market,* projected that whereas the air passenger market in the United States would grow by 1.9 percent annually from 1997 to 2016, it would expand by over 7 percent per year in the Asia-Pacific region, and by 9.6 percent per year in China.[22] The share of total world passenger capacity operated by airlines in North America would decline from 39 to 28 percent. European airlines would increase their share of total world passenger capacity from 25 to 28 percent, and Asian airlines would raise their share from 25 percent to 32 percent—giving Asia the largest share of world passenger capacity.[23]

Such growth requires larger airline fleets: Boeing estimated that Asia will generate more than a quarter of world demand for jetliners between 1995 and 2010. Sometime in the early twenty-first century, Asia will sur-

Figure 7.2   Asian Markets as a Share of World Totals, by Aircraft Types and Models

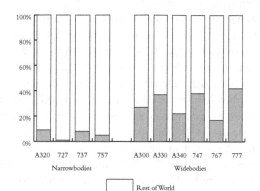

Source: Calculated from data reported in the "World Airliner Census," *Flight International* (November 18–24, 1998) pp. 44–69.

Note: Asia data includes only IATA member airlines from the Asian region, excluding the Middle East and South Asia. The A320 catagory includes he A319 and the A321 derivaives; he A300 catagory includes the A310 derivattives. Data include aircraft in service plus firm orders.

pass the United States as the world's largest commercial aircraft market.[24] Airbus forecasts that not only will Asian airlines expand their fleets to meet rising demand for air travel, but they will also increase the average size of their aircraft more rapidly than will airlines in other regions. The imperative for higher capacity aircraft in Asia stems from the congestion already affecting the region's airports. There simply are not enough landing slots to accommodate a drastic increase in the number of airplanes serving the region. Asia-Pacific airlines have a significantly higher average aircraft capacity (239 seats per plane) than any other region (159 in North America, 179 in Europe), and Airbus predicts that the Asia-Pacific average will rise to 338 seats per plane by 2016 (compared to 199 in North America and 231 in Europe).[25] Not surprisingly, wide-body aircraft are more important in Asian markets than narrow-body planes, as shown in figure 7.2. For both producers, Asia accounts for a larger share of wide-body sales than it does of narrow-body sales. Thus, the Asia-Pacific region will largely drive demand for very large aircraft, defined as planes larger than the 747, which carries about 400 passengers. Airbus expects that the Asia-Pacific region will receive 54 percent of all deliveries of "super-jumbos," the niche it is trying to fill with its six hundred seat A380.[26]

Airbus published its major market analysis, covering the 1997–2016 period, in March 1997, near the beginning of the Asian crisis. But subsequent developments did not lead the consortium to alter its broad market forecasts. In fact, at the Asian Aerospace '98 trade show in February 1998, Airbus declared that "Despite the current financial difficulties in Asian markets, Airbus Industrie (AI) . . . has not amended its current market forecast for the region, which was published in early 1997."[27] In its recent forecast for the 1999–2018 period, Airbus has reduced its projection for the rate of growth in air travel in Asia to 6.3 percent annually (Boeing estimates 6.1 percent).[28] Still, Airbus projects that Asia will need to increase seat capacity by 1.2 million, translating into forty-three hundred new planes, worth a total of $450 billion. Adam Brown, Airbus's senior vice-president of strategic planning, declared that "Airbus Industrie remains extremely optimistic about the medium- and long-term prospects for Asia air travel. We predict that the fifty-two largest airlines in the region will lead the world in terms of overall growth in capacity over the next twenty years, to the point where they will be flying as many seats as the airlines of Europe by the year 2018."[29]

Nonmarket Position

Because of the immense up-front investment required to build aircraft, Airbus could never have entered the industry without substantial support from the governments of the member companies. Governmental assistance has generally come in the form of "launch aid," consisting of subsidized loans—or, according to Boeing, grants—to pay the massive costs of design, development, and tooling for each new model. European governments have also provided export financing, guarantees against exchange rate losses, tax breaks, and debt forgiveness. Estimates of the subsidies for Airbus run to over $25 billion over the first twenty years of its existence.[30]

By the same token, U.S. aircraft companies, including Boeing, could not have remained in the industry without huge government subsidies. In fact, over its first twenty years in the commercial aircraft business, Boeing, like Airbus, lost money.[31] The U.S. government has supported the aerospace industry with R&D funding and military procurement. The spillovers from government sponsored defense R&D to commercial aircraft were frequently quite direct. The 707 was based essentially on the design of the KC-135 tanker, developed with Pentagon funding. Lockheed has sold commercial versions of its military transport aircraft. The development of the first American jet engine was financed completely by the Department of Defense. In addition, high-bypass, turbo-fan engines developed with Pentagon money for military transports made possible the 747, the L-1011, and the DC-10.[32] Profits from military sales allowed Boeing and Douglas to keep their civilian production lines open during hard times.[33]

III. Strategic and Tactical Analysis

Organizational Strategy and Tactics

The peculiar organizational structure of Airbus influences its commercial activities and its relations with governments. AI is a partnership of independent aerospace companies, organized under French law as a Groupement d'Intérêt Économique (GIE). Owned originally in equal shares by Germany's DASA and France's Aérospatiale, AI's shareholders in 1999 were DASA (37.9 percent), Aérospatiale (37.9 percent), BAe (20 percent), and Spain's Construcciones Aeronaúticas (CASA, 4.2 percent). AI owns no capital of its own, and in fact does not build airplanes. The four partners retain their separate corporate identities and their individual production facilities; they are also individually and jointly responsible up to their full capital for the obligations incurred by AI.

AI has three main roles: to arrange sales and service agreements with buyers, to oversee coordination of the dispersed technical and production activities, and to conduct flight testing and certification. Parts and sub-assemblies are produced at scores of locations throughout Europe and the rest of the world, with final assembly at plants in Hamburg and Toulouse. Airbus's advanced production software coordinates the complex flow of components. As a GIE, AI is exempt from taxation. All costs and revenues flow through the participating firms, which are each responsible for their own national taxes. Airbus is therefore not obligated to publish financial results. These are distributed among the balance sheets of the partners and mingled with their other business activities. Boeing and the U.S. government have repeatedly attacked this lack of financial transparency as a means for Airbus to obscure its financial non-viability and its resultant dependence on state subsidies.

In 1996, Airbus announced its intention to transform itself into a "normal" private corporation, with a single organizational structure. Progress toward that objective has been slow, largely because of disagreements between

Aérospatiale, which is still government controlled, and its British and German partners.[34] Finally, in mid-2000, the Airbus partners signed an agreement to create the Airbus Integrated Company (AIC). The new corporate entity, scheduled to come into being in early 2001, is to be registered in France as a simple shareholding company. What made the reorganization possible was the merger of DaimlerChrysler Aerospace, Aérospace Matra, and CASA. The merged company, called European Aerospace, Defense, and Space (EADS), will own 80 percent of the shares of the AIC. The changeover to a private corporate form would entail substantial changes, not least of which would be the obligation to report profits and losses to shareholders.[35] The new Airbus will have centralized control over purchasing and production, and will not be subject to vetoes by the partner firms.

Market Strategy and Tactics

Technical and financial requirements of the industry and the specific characteristics of Asian markets have together exercised a major impact on Airbus's strategies. In addition, Airbus must respond to the strategic moves of its competitor, Boeing. In this section, we review Airbus's market strategies, placing them in the context of Asian requirements and Boeing's activities.

Family Concept and Commonality.   The key to long-term survival for any manufacturer of commercial aircraft is the creation of a broad product line that satisfies the various needs of airlines. After the success of the 707, Boeing developed a family of aircraft ranging from the diminutive 737 to the gigantic 747. There are many advantages associated with the family concept. The most obvious is that a manufacturer can offer a variety of aircraft to satisfy the myriad needs of airlines. Depending upon the scale of its routes and market dictates, such as passenger load and frequency, a Boeing customer may select from the 717, 737, 747, 757, 767, or 777, in addition to their many derivatives.[36]

Another advantage is financial. Steady sales of the 737 and 727, the two best-selling commercial aircraft in history, helped finance development of the other members of the family,[37] and sustained the company when sales of those products lagged. Without the financial cushion provided by these two workhorses, it is all but certain that the massive development costs and anemic early sales of the 747 would have doomed Boeing.

The creation of a family, by taking advantage of the lessons learned during production and airline operations, reduces production costs.[38] Furthermore, the economies that derive from commonality of parts and subassemblies are readily apparent. For instance, the fuselage cross-section originally designed for the 707 was later utilized in the 727, 737, and 757.[39] Airbus has followed a similar path: the smaller A310 and the newer A330 and A340 share fuselage dimensions with the original A300. In the immediate aftermath of the launch of the A300, internal debate raged over whether Airbus should develop a family of airliners.[40] However, a consensus was soon reached regarding the wisdom of the family strategy that persists today. Figure 7.3 shows the combinations of range and passenger capacity available in the current Airbus family. As Vice President of Marketing Colin A. Stuart notes "We do not intend to be a niche player, and we see ourselves consistently chasing at least 50 percent of the market, with a complete family of aircraft with one hundred to one thousand seats."[41] Figure 7.3 illustrates the importance of offering a range of aircraft types, as Airbus sales really took off after the introduction of narrow-body models with A-320 in 1988.

The dynamics of the Asian market illustrate the advantages provided by a broad family of airliners. For many heavily traveled regional routes, seating capacity is paramount, whereas carriers on thinly traveled routes from Asia to Europe, North America, and the Middle East place a greater premium on range, comfort, and safety. In this regard, the "two plus four engine formula" for the A330/A340 has proven a valuable asset in the region.[42] While the two aircrafts share common wings and cockpits in addition to the A300 fuselage cross-section, differing engine configurations and seating capacities allow them to perform distinct missions.[43] The range and economy of the smaller, four-engine A340 is well suited to thin, ultralong-haul routes, while the larger A330 twin excels in high-volume regional service.[44] For instance, Cathay Pacific uses the A330/A340 as complementary aircraft for intercontinental and regional routes, since the near identical cockpits allow for cross-crew qualification.[45] Not only does this commonality enhance scheduling flexibility, it allows long-haul crews to keep skills current by practicing takeoffs and landings on shorter, regional routes.[46] Philippine Airlines cited commonality in training and subcomponents after becoming the second airline (after Cathay Pacific) to operate jointly the A330/A340 family.[47] In contrast, Singapore Airlines (SIA), while conceding that commonality in crew, training, and spares gave the A330 an early advantage over the 777, ultimately chose the Boeing product.[48]

Nevertheless, though the 767 cockpit was grafted on the nose of the larger 777, and the identical cruising speeds of the 747 and 777 allow those aircraft to be swapped without changing flight schedules, Airbus holds a

Figure 7.3   Airbus Family, March 2001

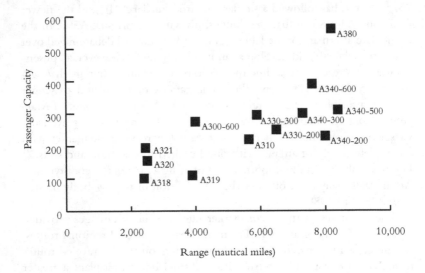

decided advantage in terms of commonality. Despite size disparities, the A330/A340 wide-body tandem, and the single-aisle A320 family share similar handling characteristics due to their analogous cockpits.[49] All Nippon Airways (ANA) selected the A340 because its side-stick controls matched those of its A320s.[50] Similarly, Dragonair (Hong Kong) leased A320s due to their cockpit commonality with its A330s.

Range.   Geography ensures that range will be an important consideration for international carriers in Asia. Destinations in Europe and the Middle East are separated from the Far East by the Eurasian landmass, and North America lies beyond the vast Pacific Ocean. Alan Pardoe, A330/A340 product manager, contends that the 747–400 is often a poor fit for Asian carriers: "About 60 percent of all 747s are purchased for their range, not capacity, and they are too big to be profitable [on very long-haul routes] on a non-stop basis."[51]

Thus A340 pitches invariably emphasize range over capacity. Print advertisements tout the craft's nonstop range ("One leg, no hopping"), and an around-the-world speed record set by an A340 was widely publicized by Airbus.[52] Ironically, the A340 has benefited from both the failures and

Figure 7.4   Deliveries, 1974–1999, by Aircraft Type

Deliveries, 1974–1999, by Aircraft Type

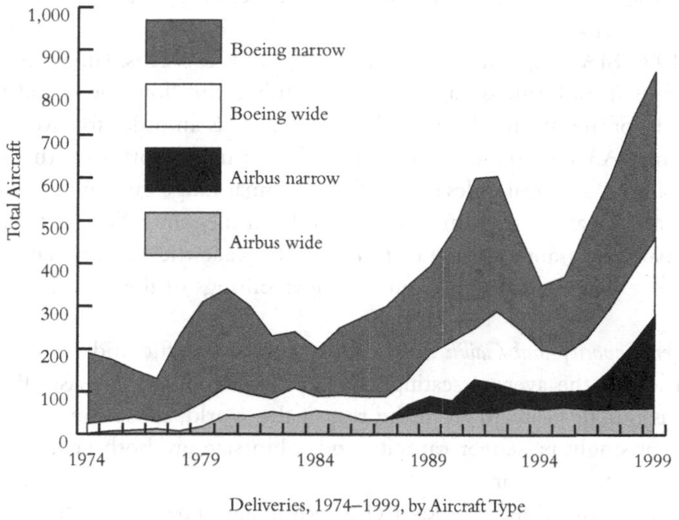

Source: For 1974–1998, Airclaims CASE Database; for 1999, Airbus Inustrie (www.airbus.com) and Boeing (www.boeing.com).

triumphs of the competing MD-11. After the MD-11 failed to meet range guarantees, SIA cancelled their order and purchased the A340 instead.[53] But Cathay Pacific ordered the A340 to counter their competition's non-stop MD-11 flights.[54] Future derivatives of the A340 will also emphasize range. The A340–500 will retain the capacity of the current A340–300 while increasing range by one thousand nautical miles.[55] With its eighteen hour range allowing nonstop flights between, for example, Hong Kong and the eastern coast of the United States, as well as Australia and Europe, the A340–500 will be configured with lower deck beds for high yield passengers.[56] Encroaching on the 747–400's market position, the stretched A340–600 will boost capacity to 378 in a three-class layout, but will retain the current model's range.[57]

By contrast, the larger A330 has been unable to match both the size and range of the 777. Cathay Pacific selected the Boeing offering largely because the A330 lacked the range for flights to Bahrain, Dubai, and Sydney.[58] "The big problem we saw with the stretched A330," according to Cathay Pacific engineering director Stewart M. John, "was that although it gave us fifty more seats, it didn't give us the range flexibility we need."[59] Cathay Pacific preferred to wait until the stretched 777–300 would ultimately be available.[60]

In 1999 SIA swapped its entire fleet of A340s for 777s, largely due to complaints from business passengers who dislike the slow speed and narrow cabin of the Airbus plane. While SIA still holds an order for five ultra-long-range A340–500s, sources within the airline report that the new plane falls 400 nautical miles short of the original range guarantee because it is six and a half metric tons overweight. Ironically, these failures to meet performance guarantees could help the 777 displace the A340 much as the A340 once benefited from the similar shortcomings of the MD-11.[61]

Passenger Capacity and Cabin Size.   Due to heavy traffic and constrained infrastructure, the average seating capacity of airliners in the Asia-Pacific region is much greater than in the rest of the world, as seen in table 7.3. Not surprisingly, passenger capacity and cabin size are both critical considerations for Asian carriers.

In this regard, Boeing holds a clear advantage. The passenger capacity and fuselage diameter of the 747–400 is unrivaled. Plus, its nearest competitor, the 777, is also a Boeing product. With a cabin width only ten inches less than the 747–400, the 777 allows one or two more economy seats per row, in addition to an extra business or first class seat per row.[62] Three Asian carriers—Cathay Pacific, Japan Airlines (JAL), and ANA—were members of the "Gang of Eight" that helped Boeing design the 777. [63] Malaysia Airlines System (MAS) chose the 777 over the A340 despite its higher price because its larger cabin would generate more revenue.[64] IA managing director Cheong Choong Kong proffered a similar explanation for choosing the 777 over the A330.[65]

Attempting to neutralize this disadvantage, Airbus has emphasized range and the greater comfort afforded by its cabin layout. The latter point is highlighted by two print advertisements for the A340, one noting that economy class passengers are "never more than one seat from the aisle," and the other stressing that "there is no middle seat in business class." However, seasoned business travelers report that business class is rarely full and that the larger Boeing jets offer more headroom for passengers in window seats.[66]

Another view holds that size will prove increasingly irrelevant as airlines vie for coveted, high-yield, first-class, and business–class passengers in Asia: "You try to get the backpackers . . . into someone else's aircraft," notes David Jennings, Airbus vice president for marketing.[67] Accordingly, Cathay Pacific uses A340s configured for 261 passengers in a three-class layout for

Table 7.3 Average Number of Seats per Aircraft

Average Seating Capacity	1998	2018 (projected)
Asia-Pacific	244	310
World	180	218

Source: Airbus Industrie (1999).

point-to-point service, deploying its 343 seat, two-class 777s in regional routes.[68] Cathay Pacific maintains that the 777 is the wrong size for long-haul routes since its seating capacity is too close to that of the 747–400.[69]

Yet the Asian market size often trumps other considerations. For SIA, the original A340–200 possessed sufficient range, but lacked the requisite payload.[70] The airline later purchased the A340–300 model enhanced with additional payload capacity. Korean Air, while interested in long-range aircraft, was not interested in purchasing the A340 because, as stated by President Yang Ho Cho, "We need bigger aircraft."[71] Thus, notwithstanding Jennings' earlier comment, Airbus Chief Operating Officer Volker von Tein states: "Today our efforts are almost entirely devoted to our high capacity aircraft product range."[72] Thus Airbus is producing stretched versions of the A330 and A340.

The centerpiece of the Airbus strategy on the capacity dimension is the A380, launched at the end of 2000, a double-decked giant whose capacity and technology eclipse those of Boeing's graying 747–400. The A380 is capable of carrying as many as twelve hundred passengers. Of the nine airlines who originally expressed interest in the A380, five—JAL, SIA, ANA, Cathay Pacific, and Korean Air—are Asian.[73] Adam Brown, Airbus's senior vice-president of strategic planning, stated that the original impetus for the A380 came from Asia: "The market is extremely concentrated. Demand for aircraft with more than five hundred seats will be very largely driven by the airlines of the dynamic Asia-Pacific region, which will account for about sixty-five percent of all deliveries of these very large aircraft."[74] In 1999, Airbus estimated that of the ten routes best suited to the A380, eight originate in Asia, and that six of the top ten A380 airports are located in Asia.[75]

Operating Costs. Another important consideration for airlines, and thus manufacturers, is operating costs. The A380 must deliver operating costs 15–20 percent lower than the 747–400 to prove economically viable to airlines. Even so, these savings will prove illusory if the colossal jetliner is not filled on a regular basis.[76] Both manufacturers claim superior economies for their existing offerings. Boeing claims that 777 operating costs are lower than the A340 and MD–11, while still competitive with the A330.[77] Brown rejoins that the A330–300 has "greater payload/range capability, and substantially lower direct operating costs per seat than the 767–300ER."[78]

Less biased assessments are available from the airlines. In most regards, Airbus performs well in this area. According to Cathay Pacific chairman Peter Sutch, its A340s will prove less expensive than 777s due to lower engine maintenance costs, acquisition costs, and landing fees, all the result of its lower weight. "At this stage if you want to be as risk-averse as possible, then the A340 is the way to go. The economies of the Cathay Pacific route structure say the A340 is the safer way to go."[79] In addition, Cathay Pacific's new A330s boast operating costs far superior to those of its oversized 747–200s.[80] SIA's evaluation of the A340–300 is comparable: "We rate the aircraft highly in terms of economy, modernity, and comfort," notes Managing Director Cheong Choong Kong.[81] Yet despite the A330's commonality with its existing A340 fleet, SIA purchased the 777 due to its lower acquisition and operating costs, not to mention its superior performance.[82]

After-Sales Service. While operating costs have generally proven reasonable, Airbus Chairman Jean Pierson conceded that "high spare parts are an issue we will solve. European component manufacturers must increase their [industrial] productivity. Their customer service sometimes needs to be improved. We plan to replace inefficient vendors."[83] Also, Northwest Airlines complained about the high price of A340 computer spares, and added that the suggested daily maintenance requirement to shut down computers and reboot were an "unacceptable burden" for the airline.[84]

However a consensus exists that the after-sales support provided by Airbus has improved drastically since the consortium's infancy. As Reinhardt Abraham, Lufthansa's deputy managing director notes, "Airbus's after-sales service was inferior to the Americans' for a long time. We all had to work very hard to speed them up in this respect, and I have to say that they have improved their field support and field organization. They have also improved to some extent as far as engineering support is concerned."[85] Airbus has also attained a high level of dispatch reliability in the developing world, a consequence of its far-ranging support services. In India, for example, Airbus established a bonded warehouse in Bombay to allow rapid access to spares for Air India's A310s.[86] Later, the debut of Indian Airlines' A320s was accompanied by an Airbus-sponsored support program.[87] Furthermore, Airbus test pilots were later flown to India to boost utilization rates of those A320s.[88]

Two print advertisements tout the responsive customer support provided by Airbus. Airbus's resident customer support managers represent over twenty-five nationalities and "can be relied upon to understand your country's culture," while the "Customized Lead Time Programme" cuts operating costs through a customized delivery schedule that ships parts within two hours while reducing inventory by a claimed 60 percent. Nevertheless, many Asian carriers are very conservative and famously risk-averse, and thus prefer to maintain their own stockpiles of spares.[89]

Safety. A300 sales initially suffered due to a widespread perception of twin-engine aircraft as more dangerous.[90] This apprehension has mostly dissipated due to the stellar safety records posted by the A300/A310 and the 767. However, with the four-engine A340 locked in a tight sales battle with the 777 twin, Airbus now has little incentive to soothe flyers' fears about twins. In fact, a two-page print advertisement for the A340 features a head-on view of the aircraft with the four engines prominently displayed, and the suggestion that pilots prefer four engines rather than two. ANA selected the A340 for long-range, overseas flights, because its managers thought it was safer than the 777, describing the four-engine configuration as the "tie breaker."[91] But Airbus sells twin-engine planes as well, and will sometimes dramatically demonstrate the safety of the configuration to reassure skittish customers. Thus, the A330 twin, as part of an evaluation by Chinese airline officials, successfully performed engine-out tests in the high-altitude of Lhasa, Tibet.[92]

Advanced Technology. Since its inception, Airbus has employed advanced technology to distinguish itself from its more conservative rivals. According to an unnamed European company manager: "There was general agreement many years ago that we in Europe had to have the dual approach. We needed the technology to make the aircraft efficient and attractive, and we needed a sufficiently broad product range to remain viable despite the ups and downs of the various market segments in terms of size and type of aircraft."[93] Airbus pioneered cockpit automation (including the controversial side-stick controllers), fly-by-wire technology, and the use of composite materials. When the two latter innovations were later included in the rival 777, Airbus could hardly contain its glee at the about-face made by the naysayers at Boeing. Much earlier an unnamed Airbus manager gloated:

Several years ago, our U.S. competitor told a lot of scare stories on how risky Europe was playing it by deciding to use the horizontal tail-mounted fuel tank on the A310–300, but now Boeing is offering this same feature for its own products. We developed the minimum two-crew cockpit first, and Boeing had to rush to introduce it in Seattle; we took the step of introducing wing-tip fences on our aircraft, and went to the side stick controllers in the A320 first. I guess you can say the tables have been turned when it comes to new technology.[94]

Advanced technology will prove even more critical to the success or failure of the outsized A380, designed to fit in an eighty meter by eighty meter box to facilitate operations in existing airports. Wake vortex characteristics must be carefully designed to maintain current separation distances between the A380 and smaller aircraft. Increasingly stricter noise regulations have forced Airbus to design in noise performance comparable to that of the much smaller A340. A new composite material, the multi-ply laminate of aluminum and glass fiber dubbed Glare, will likely be employed to improve flame resistance without increasing weight.[95]

Availability and Speed of Delivery.   For airlines, speed of delivery has been a critical consideration due to the cyclical, highly unpredictable nature of industry demand. Paradoxically, a competitive advantage for Airbus has been its relative availability compared with more popular Boeing models. A 1978 sale of A300s to Pakistan International Airlines was clinched because no Boeing products were available until 1984 or 1985.[96] Similarly, according to SIA, the A300 was chosen because the 757, 767, and the A310 derivative were all "out of the running because they were not available until 1983 at the earliest."[97]

More recently, airline commitments have been guided by early launch decisions. Cathay Pacific selected the A330/A340 largely due to its immediate availability.[98] Conversely, because JAL "needed to have the guarantee of aircraft availability when we want them," the formally launched 777 bested proposals for the A330 and the stillborn MD-12X.[99]

Both manufacturers have sought to improve production efficiency in order to speed lead times. Airbus has implemented "discontinuous manufacturing," an innovation that will tie production rates to firm orders to improve flexibility, speed of delivery, and cost-competitiveness.[100] Production cycles are slated to drop from twelve to nine months for narrow-body jets, and fifteen to nine months for wide-bodies with a long-term goal of a mere six months from customer specification to delivery. As Aérospatiale Managing Director Yves Michot explained: "Shorter production cycles are the right answer to airlines' insistent demand for late decisions on airframe customization."[101] Boeing has already slashed its eighteen-month lead time in the early 1990s to ten months and ten-and-a-half months for narrow-bodies and wide-bodies, respectively.[102]

Pricing/Discounting/Financing. As *Aviation Week* observed: "All other selling points, quality, lead times, equipment options, commonality, financing, and price being equal, it's the latter that carries the most weight among airlines."[103] In an effort to gain market share, Airbus has continually and significantly discounted its products. An early Thai International contract for A300–600s unabashedly acknowledges a discount of $10 million per plane.[104] This pricing pressure has frequently aroused the ire of its American competitors, with McDonnell Douglas complaining to Congress that the pricing of the A340 relative to the MD-11 is impossible to explain in commercial terms.[105]

Airlines play the manufacturers against one another in order to keep prices low. In addition to the cut-rate prices offered due to global recession, Cathay Pacific selected Airbus as a supplier to "keep Boeing honest."[106] Today, price cuts as much as 35 percent to 40 percent per aircraft are common. The 737 and A320, with list prices approximating $35 million, can be purchased for less than $25 million.[107]

Financing terms can sometimes prove decisive when airlines evaluate competing aircraft. In Asia, the well-regarded 727 faced stiff competition because the upstart A300 offered "much better terms."[108] Generous financing may even persuade airlines to purchase aircraft that are ill-suited to their requirements. One Lockheed official charged that "under these circumstances, they will sometimes buy the A300 and rearrange their fleet and their routes to accommodate the financing."[109]

Also controversial are "trade-in" or "buyback" plans where one manufacturer agrees to purchase or guarantee the future sale of a rival manufacturer's aircraft. To secure an order for ten 777s, Boeing agreed to purchase SIA's entire fleet of seventeen A340–300s if the airline is unable to dispose of them first.[110] An Airbus spokeswoman criticized the arrangement as "a new kind of price war. Instead of using price tags as the ammunition, (Boeing) is using planes." In retaliation, Airbus threatened to withhold product support for the remarketed planes by refusing to guarantee performance, honor warranties, or provide parts and service. A Boeing executive, claiming such trade-ins are common practice in the industry, condemned Airbus's threatened retaliation as "irresponsible" and potentially dangerous to the flying public.[111]

Government subsidization of financing is a continuous point of dispute among manufacturers. In the United States, the role of the Export-Import Bank has been expanded to meet the perceived foreign challenge, according to Vice Chairman William F. Ryan. Referring to the purchase of 767–300s, he stated: "With foreign export credit agency financing available in support of the Airbus bid, Ex Im Bank's [the U.S. Export-Import Bank's] preliminary commitment to ANA was essential to assure that Boeing could compete fairly with respect to financing for this order."[112] In General Agreement on Tariffs and Trade (GATT) negotiations, financing has been a central issue.

China's initial interest in the A300/A310 followed $5 billion in British export credits.[113] Moreover, the first export loan from a European bank to Japan financed Toa Domestic Airlines's (TDA) purchase of the A300.[114] A more recent initiative is the Airbus Finance Company, described by an official as the "right working tool to support commercially strong but financially weak airlines, a trend that is not expected to reverse in the future."[115]

Nonmarket Strategy and Tactics

While many Airbus sales in Asia are attributable to market strategies, some sales can be directly or indirectly linked to the nonmarket strategies employed by the consortium. Nonmarket strategies include subcontracting and, less frequently, co-production offers: investment in local training or maintenance facilities; strategic manipulation of forecasts; diplomatic pressure applied by institutions such as the European Community (EC), European leaders and trade representatives, lobbying, and even bribery of airline officials.

Subcontracting and Co-Production.   Many Asian airlines are fully or partially state-owned. Thus, Airbus has often attempted to promote purchases by offering national aerospace companies lucrative subcontracting deals. While subcontracting is often profitable, the major benefits accruing to the subcontractor are the technology transferred, and the learning that occurs during production. Indeed, from the point of view of many Asian aerospace companies and their governments, technology transfer is a high priority. Technology transfer thus becomes a *quid pro quo* for market access.

India, soliciting proposals for medium-capacity, long-haul airliners, demanded that 30 percent of the winning aircraft's content be locally produced. While such a demand may seem excessive, offset requirements that explicitly link local subcontracting with airframe purchases are common in China, Japan, South Korea, and Indonesia.[116] An Indonesian offset requirement mandated that Airbus must first enter into business with the state-owned IPTN aerospace firm before Garuda Indonesia, the state-owned flag carrier, would purchase Airbus products. Soon after, an Airbus delegation arrived to provide technical advice for the Japanese Aviation Administration and the U.S. Federal Aviation Administration certification of IPTN's N250 turboprop commuter airliner.[117] Airbus test pilots helped develop a flight test program for the N250, and Airbus partner Daimler Benz provided manufacturing advice. In addition, IPTN was awarded a contract for at least 200 shipsets of A330/A340 flap-track cartridges for DASA.[118]

While such linkages are somewhat less explicit in China, parceling out subcontracting jobs is no less important. Thus, Xian Aircraft Corporation produces wide-body access doors for Aérospatiale; the Shenyang Aircraft Corporation is the single source supplier for emergency exit doors for the A319 and A320; and Guizhou Aviation was named a supplier of ground support equipment to maintain A320 and A330/A340 transports.[119] As late as 1993 only five Airbus transports were flown by Chinese airlines; by 1998 the equivalent figure was fifty-three, with many more on order or option.[120]

Efforts to offer subcontracts are not always successful. An industrial partnership was established with Malaysia Airlines Systems (MAS) for the manufacture of components, with Airbus providing technical assistance, training, and specialized equipment.[121] MAS also builds seats for BAe as "part of a broader effort by Airbus partner companies to secure orders from MAS." Despite these relationships, plus an offer to become sole supplier of A340 composite cargo liners, MAS chose the 777.[122]

In the long run, however, Airbus is skeptical that subcontracting, the long-preferred strategy of Boeing, will prove sufficient to placate the industrial ambitions of Asian nations. Upon offering Japan partnership in the A380, Adam Brown (Airbus's senior vice-president of strategic planning) remarked: "I raise the question as to how much longer Boeing can hope with such an approach to satisfy the legitimate aspirations of the dynamic developing nations in the Asia-Pacific region."[123] To that end, in 1993 Airbus Managing Director Jean Pierson visited the three largest Japanese aerospace contractors, Mitsubishi, Kawasaki, and Fuji, to discuss collaboration on the A380.[124] The Japanese resisted Pierson's entreaties, and to this point have shied away from collaboration on the A380. In fact, Japan was offered joint development of the SA-1 and SA-2 130- to 150-seat transports (later to become the successful A320 family) and declined. In 1996, China also rejected Airbus' offer of partnership on the A380.[125]

Such reticence can be explained partially by the substantial financial risks involved with the development of any commercial aircraft, especially the massive A380. But the recently cancelled "Asian Express," a joint venture between Airbus, Italian contractor Alenia, Singapore, and AVIC provides a cautionary tale. Having already produced licensed versions of McDonnell Douglas MD-80/MD-90, AVIC was eager to jointly develop and produce a 95–115 seat regional jet. However, the proposed AE316/AE317 was shelved when Airbus began to question its cost-efficiency and Singapore doubted its viability. Instead, for a mere $300 million, Airbus chose to develop the 106-seat A318, essentially an A319 truncated by five fuselage frames, while promising China participation in future endeavors.[126] Contracts signed at the 1999 Paris Air Show for Xian Aircraft to produce 500 A320 wing shipsets, and guaranteeing the participation of AVIC engineers in the A318 program, are intended as "compensation" for the collapse of the earlier, more extensive collaborative venture.[127]

Investment in local training and maintenance facilities has often been hampered by considerations other than efficiency. A 1982 *Aviation Week and Space Technology* article noted that joint A300 maintenance efforts were hindered by national pride, differing management strategies, politics, and competition. The establishment of a spares sub-depot in Hong Kong and a training center with SIA provided the foundation of a broad support network for Asian carriers that Airbus has continually sought to upgrade.[128]

Although Airbus had maintained an office in Beijing since 1990, the $170 million Hua-Ou Aviation Training Center was intended to symbolize Airbus's commitment to China, and quickly paid dividends when China announced an order for thirty A320s less than a year after groundbreaking. Airbus contributed $70 million, while China supplied land and personnel. The training center is a grand gesture akin to McDonnell Douglas's establishment of a MD-80/90 production line in Shanghai.

According to Airbus China President Rolf Rue:

What the Chinese want to see from us is a full commitment to the support of a fleet of aircraft in China. We wanted to improve that support, and the Chinese wanted to see an improvement. It was a commitment we made independent of any orders. It was important, but not directly linked . . . The package of services that comes with our aircraft is almost as important to the operators as the aircraft itself. By building the center in Beijing, we are able to demonstrate to the airlines that the support is there when they need it. This allows them to concentrate on the technical and commercial advantages of our products.[129]

The training center, a joint venture with China Aviation Supply Corporation (CASC), will be turned over to the CASC in 2005.[130] Parts suppliers will also be based in the center.[131]

The Beijing training center is noteworthy not only due to its scale, but also because the only comparable facility outside France is located in Miami. But, as Premier Jiang Zemin told Boeing, what the Chinese want is "training, training, training." Subsequently, Boeing gave two 737 simulators to the Civil Aviation Authority of China's Civil Aviation Flying College in Chengdu, improved its training throughout China, and opened a 24-hour spares center for itself and eighteen suppliers at the Beijing Airport.[132]

Airbus suppliers have followed the consortium's lead, with French contractors Sogema, Sfena, and Jaegar establishing support facilities in Singapore in 1981.[133] Later, in a joint venture with Thai Airways, Sogema invested $80 million to establish a regional Airbus maintenance center in Khon Kaen, Thailand.[134] Even Air France, in what was described as a "marketing coup for itself and Airbus Industrie," leased five A320s to Vietnam Airlines while providing flight crew and technical training.[135]

Strategic Use of Market Forecasts. Another nonmarket strategy is the employment of market demand forecasts as a tool to obtain sales or impede competitors. Airbus predicts that airlines will require 1,442 aircraft larger than the 747–400 over the next twenty years.[136] In contrast, Boeing forecasts a demand for 480 superjumbos. Boeing contends that there will be insufficient need for a superjumbo, pointing out that 80 percent of the growth in available seat-kilometers over the past decade has been due to increasing the number of flights, as opposed to the use of larger airplanes.[137] Airbus officials claim that Boeing has "cynically manipulated its own forecast in a transparent attempt to make Airbus appear the 'odd man out,' and to discredit Airbus Industrie's forecast of demand for very large aircraft."[138] Adam Brown, Senior Vice-President of strategic planning, calls the Boeing forecast "a manipulation, an attempt to dissuade Airbus from entering their market, to eliminate their [high capacity aircraft] monopoly." Pierson echoed this sentiment: "Boeing wants to protect the 747–400's monopoly in the market's upper segment. They intend to achieve this with limited—or no—additional investment. But they will have to revise their plan when the A380 is in full production and has obtained orders."[139] Estimates of Boeing's profit on each 747–400 run as high as $45 million.[140] Considering this, a strong incentive exists for Boeing to discourage its rival's entry into that market segment by minimizing the potential reward.[141]

Boeing has continually employed pessimistic forecasts of demand in an effort to discourage its European nemesis. Initially arguing that airlines would never accept the large A300 twin, Boeing was forced to respond to its success with the 767. After denying a market existed for the advanced technology A320, Boeing was later obliged to respond with the next-generation 737. Faced with the prospect of the A330/A340, Boeing first insisted that a combination of 747s and 767s could satisfy the needs of airlines, but eventually countered with the 777.[142]

Diplomatic Pressure.   Since Airbus competes in a highly visible, important trade sector, and is a consortium composed of nationally owned aerospace companies, it should be no surprise that diplomatic pressure, often at the highest levels, is utilized to improve Airbus's fortunes. In addition, the French government has often linked inducements such as landing rights, technical assistance, and special trade agreements to the purchase of Airbus transports.

In 1976, Boeing VP of sales Clarence V. Wilde complained that Airbus had the advantage of a "real government-to-government-type selling effort" after India received French technical support for powerplant development and tariff concessions for textiles in conjunction with its purchase of the A300B. Similar concessions allegedly accompanied the sale of the A300B to Korean Airlines (KAL).[143] Before the House Banking international subcommittee, Boeing treasurer J. B. L. Pierce testified:

> We can compete with Airbus and other European aircraft manufacturers on cost and technical merits, but we cannot compete with the national treasuries of France and Germany and other European countries . . . Both of these financing programs (long-term debt financing and long-term lease financing) are superior to those which U.S. manufacturers can offer, even with Export-Import Bank assistance. Further, Airbus Industrie, because of multi-government ownership, has been able to stimulate and secure the support of various government ministries to add further momentum to the sales campaigns. This has enabled Airbus Industrie to offer flexible and imaginative financing programs and, if necessary, link other inducements, such as landing rights, to the sale of their aircraft.[144]

A Treasury Department inquiry followed a California Congressman's allegation that China Airlines was offered landing rights in Paris to pick the A300 over the DC-10.[145] Though the United States could sometimes match financing terms through the Export-Import Bank, one official conceded the bank's broader impotence: "The problem has been the perks that come with buying the Airbus such as landing rights, free training, and so forth that they sweeten the deal with. Our hands are tied as far as those are concerned."[146] Yet such a strategy could sometimes backfire. After the British resisted expanding Hong Kong landing rights, Thai Airlines reduced its A300 order.[147]

As competitive pressures increased throughout the 1970s, American firms sought increased participation in trade policymaking. Manufacturers of commercial aircraft were no exception and, catalyzed by the exceptionally generous terms that accompanied the sale of A300s to Eastern Airlines, pressed for a sectoral agreement.[148] While the resulting 1979 GATT Agreement on Trade in Civil Aircraft "successfully eliminated most traditional trade impediments and substantially liberalized aircraft trade," it nonetheless failed to tackle the fundamental dispute of industrial policy support for Airbus.[149] The American delegation attempted to enshrine "commercial competition" as the guiding principle of the agreement. But France and Great Britain (the EC was relegated to a formal role) insisted that the preamble to the agreement should endorse the "provision of fair and equal competition between domestic and imported products."[150] Due to European reticence, any restrictions on government support were "limited in scope, vague in stipulation, and weak in enforcement."[151]

Controversial sales to Pan Am and Air India spawned accusations that predatory pricing by Airbus violated Article Six of the 1979 agreement. After the United States threatened draconian Section 301 action against Europe, the two parties embarked upon a series of bilateral talks beginning in 1986. The Gellman Report, commissioned by the Department of Commerce, laid out the American case against Airbus. Government support to Airbus through 1989 totaled $13.5 billion, the equivalent of nearly $26 billion at private sector borrowing rates.[152] Not to be outdone, the EC retained the American law firm of Arnold and Porter to document direct and indirect government support for Boeing and McDonnell Douglas. According to the Arnold and Porter report, support from NASA, the Department of Defense, and the favorable tax code was estimated at $33.5 to $41.5 billion over the previous fifteen years.[153]

In 1987, the June launch of the A330/A340 program was followed in July by Congressional hearings that "were quite confrontational toward Airbus in tone and content."[154] The tensions reached a climax in 1988 when Daimler-Benz agreed to purchase MBB, the owner of Deutsche Airbus. While troubled by the German government's plans to absolve MBB's public indebtedness and acquire a 20 percent stake in a holding company that would oversee its interests in Airbus, what truly enraged

American negotiators were guarantees that Daimler-Benz would be re-imbursed for losses resulting from shifts in foreign exchange rates. The "forex guarantee," according to Thornton, was perceived by the United States as "an unambiguous example of government assuming commercial risk."[155] After lodging a formal complaint, a GATT panel eventually ruled in favor of the United States in 1992.[156]

European negotiators held fast to their conviction that launch aid was justified in order to remedy "competitive inequity," but eventually agreed to reduce and cap these subsidies.[157] Following much political posturing and contentious debate, a July 1992 Agreement limited direct government support to 33 percent of the total cost of new programs (existing programs were exempted), and instituted new dispute resolution procedures.[158]

Another area of contention is the lopsided composition of Japanese airline fleets. Bernard Salzer, chairman of a European Parliament delegation, raised the question, "If Airbus has achieved a 30 percent share of the world civil aircraft market, why is the corresponding share in Japan only 10 percent, and why has Rolls Royce sold no aircraft engines in Japan since 1970?"[159] The EC has pressured Japan to buy more Airbuses, complaining that JAL had never purchased Airbus offerings and had purchased the 737–400 without even considering the A320.[160] Also, ANA cancelled Airbus, but not Boeing, orders.[161]

In the aftermath of a $6 billion purchase of American planes by Saudia (the Saudi Arabian flag carrier, now called Saudi Arabian Airlines), Airbus fumed about the distorting influence of generous Ex-Im bank financing, coupled with the restructuring of a multi-billion dollar Saudia debt for military hardware.[162] After former U.S. President Bill Clinton phoned King Fahd to clinch the huge order for Boeing and McDonnell Douglas, Pierson speculated the deal was linked to a hidden agreement between the two countries, possibly regarding Bosnia.[163] In the aftermath of the Saudi Arabia deal, an anonymous Airbus official asserted: "We are a European consortium. And there is no Mr. Europe who could do a similar job to President Clinton's. We never asked for and we never obtained political support to help us in the market. And certainly not from French President François Mitterrand or from French Prime Minister Edouard Balladur."[164]

However, the above statement is wholly disingenuous. In 1997, China signed a large order for Airbus in the presence of French President Jacques Chirac during his visit.[165] In addition, while there is certainly no "Mr. Europe," Airbus's multinational framework allows Airbus a malleable national identity and the choice of various high-level representatives. Thus, when France's sale of Mirage fighters to Taiwan antagonized China, the Chinese were able to save face and order "German" A340s during the visit of Chancellor Helmut Kohl.[166] Furthermore, the Commission of the European Union (EU) represents the EU and its member countries in all multilateral trade talks. Because Airbus is the European "champion" in commercial aircraft, the EU acts as its advocate in international negotiations. In fact, as mentioned above, the EC has often stepped into the fray on behalf of Airbus. Recently, its behavior has grown more assertive, threatening to block Boeing's merger with McDonnell Douglas, and vigorously protesting its unprecedented exclusive-supplier agreements with American, Continental, and Delta Airlines.

Airbus has always employed carefully targeted lobbying as a selling tool. A high-level German trade delegation visited China just prior to the inception of Airbus China.[167] Another European trade mission to China included Airbus.[168] But Airbus soon learned that cultivating relationships with individual Chinese airlines was as important as lobbying central bureaucracies, like the Civil Aviation Authority of China (CAAC) and China Aviation Supply Corporation (CASC).[169] After negotiating solely with the central purchasing authority, Airbus was twice rebuffed when China Southern picked the 777 over the A330, and later refused delivery of A340s ordered by the CAAC.[170]

Accusations of bribery have often swirled about Airbus sales. After an Indian Airlines A320 crash, the new Indian government reiterated existing allegations of bribery by its predecessors. In Sri Lanka, a government probe accused officials of receiving "commissions" after committing to the A340.[171] Yet the political geographic scope of these charges is not limited to developing nations or Asia. Air Canada's purchase of the A320 was tainted by insinuations of bribery and spawned an ultimately inconclusive probe.[172]

It is often difficult, if not impossible, to ascertain the role of diplomatic pressure in closing a deal for airliners. China Airlines, after concluding a nonbinding letter of intent for 777s, instead chose A340s in a purchase seen by some as a rebuke of the "one-China policy" espoused by the United States. But one analyst stressed that the switch was just as likely spurred by Airbus's "making an offer China Airlines just couldn't refuse."[173]

IV. Conclusion

Airbus's strategies in Asia are formulated in an environment shaped by three forces: the technical constraints of building modern aircraft, the demands of world and regional markets, and the structure and objectives of the consortium itself. These fit quite neatly into the integrated strategy analysis presented in chapter one of this volume.[174] The market for commercial aircraft is a genuinely global one, in which various regions play differing roles. The positional analysis for Airbus focused on its status as the challenger to the dominant firm, on the consortium's peculiar organizational structure, and on the opportunities and constraints offered by its nonmarket (political) context (that is, multiple national governments plus the EU). Strategic analysis includes Airbus's arena strategy and nonmarket strategy (which we address in more detail below), and its organizational strategy. The latter element touches directly on the proposed reorganization of Airbus as a regular corporate entity. Though the restructuring has yet to take place, it is clearly seen by many in the Airbus community as a crucial step to ensure the group's future ability to compete with flexibility and agility. Tactical analysis shows Airbus making shrewd moves in the marketplace while seeking to satisfy the desires of Asian governments for subcontracting, technology transfer, and training. On the organizational side, Airbus has been willing to contract with Asian suppliers, and has on some occasions offered closer (i.e., stake-holding) partnerships in specific aircraft, so far without takers.

A more fine-grained look at one piece of the broad analytic schema might be a useful way of viewing the interplay of market and nonmarket strategies. Market strategies are at the center of the framework. We conclude by mapping some specific Airbus strategies and tactics onto the four arenas of hyper-competitive markets, as a means to explain why Airbus has been notably successful in the intensely competitive arena of commercial aircraft.

Richard D'Aveni argues that the general strategy of a successful firm in hyper-competitive markets is not to sustain an existing advantage in some sort of equilibrium, but rather to provoke a series of disruptions that offset the competitive advantages of its rival. One way of looking at Airbus's entire existence is as a continual series of disruptions aimed at creating openings for Airbus to enter markets dominated by well-established competitors, Boeing in particular. These strategic moves have involved both market and nonmarket elements. Though Airbus's gambits have not always paid off, they have scored sufficiently often to win about a third of the world market for airliners, and to position Airbus as a serious threat to Boeing's dominance.

The first arena, cost and quality, has been a crucial battleground for Airbus. Though Boeing obviously has similar interests, its dominant position probably generated some complacency in the early stages of the competition with Airbus. Airbus certainly took extraordinary measures to please the airlines that started buying its planes, including lending aircraft to Eastern Airlines (the first U.S. customer) and lending pilots to Air India.

The second arena, that of timing and know-how, has particular relevance to Airbus. Airbus has constantly sought to introduce new classes of aircraft and new aviation technologies as a means of transforming aircraft markets in its favor. With respect to timing, Airbus offers two clear examples. By being the first to identify a market opening for wide-body airplanes with two engines, Airbus undercut both Lockheed and McDonnell Douglas with their less efficient three-engine wide-bodies, and forced Boeing to respond. More recently, Airbus's planned A380 seems clearly to be an effort to identify, if not create, a market need for "superjumbos." The A380 would wipe out Boeing's monopoly on high-capacity aircraft and position Airbus to meet Asia's need, given limited supply of landing slots at its airports, to carry more passengers per airplane. In terms of know-how, Airbus has managed to pull off a series of technical innovations that Boeing (and others) initially predicted would flop in the market. The innovations, later adopted by Boeing, included the two-pilot cockpit, "fly by wire" (electronic) controls, and extensive use of composite materials.

Strongholds have been a crucial part of Airbus strategies as well. Particularly during the lean early years, Airbus could count on orders from the flag carriers of its members' home countries. However, Europe was never a genuine regional stronghold for Airbus. Boeing and McDonnell Douglas already dominated European aircraft markets when the new consortium came into being. Also, British Airways continued to prefer American-made planes for many years after the launch of Airbus. Hence it was indispensable for Airbus to attack Boeing's stronghold in the American market, which many observers thought would be impossible to crack. Thus Asia became the critical region. Though American planes dominated Asian markets initially, it was a region where no aircraft maker would be able through political mechanisms to create a stronghold. Thus we have highlighted Airbus's use of market and nonmarket strategies to succeed in Asia. On the market side, Airbus has clearly sought to produce aircraft that would meet the needs of Asian carriers for efficient, high-capacity planes. The A380 is only the clearest instance. The development of the A380, which has involved extensive consultations with Asian airlines and the enlisting of Asian subcontractors, has been a powerful signal of Airbus's commitment to Asia, and of its intention to disrupt the market at the high-capacity end. On the nonmarket side, Airbus has entered into subcontracting relationships when this was important to governments whose airlines were in the market to buy aircraft. Furthermore, European governments have been willing to offer technical and industrial assistance and development projects to countries whose airlines chose Airbus.

Finally, deep pockets have played an important role in the aircraft industry, though not in the conventional sense. In the case of aircraft, the deep pockets have been those of various home governments. Airbus's reliance on "launch aid" from national governments brought to a head the long dispute over subsidies, and led to bilateral agreement on a rule that limited government subsidies to 30 percent of the costs of development. Creative export financing also forced Boeing and the U.S. government to alter the ways in which they promoted foreign sales.

Regardless of the scheme one uses to parse the elements of Airbus achievements, the story is straightforward: Airbus, starting from nothing, took on the giants of an intensely competitive industry. But not every David conquers the Goliath. What makes Airbus a compelling case of successful strategies is that, often enough, it won.

Notes

1. See Wright (1984), pp. 18, 24; and McGuire (1997), p. 49.
2. Aircraft markets in South America, Africa, the Middle East, and Oceania are too small to make a decisive difference in Airbus sales, though significant orders from some regional airlines (for instance, Kuwait Airlines, Saudi Arabian Airlines, and Egypt Air) have arrived at critical moments in the past.
3. Newhouse (1982).
4. Mowery and Rosenberg (1989), p. 170.
5. Hayward (1994), p. 10.
6. Mowery and Rosenberg (1989), p. 172.
7. *Aviation Week and Space Technology,* July 21, 1997, p. 15.
8. Tyson (1992), p. 167.
9. *Aviation Week and Space Technology,* November 23, 1998, p. 38.
10. Tyson (1992), pp. 165, 167.
11. McIntyre (1992), pp. 35, 36.
12. U.S. Civil Aviation Manufacturing Industry Panel (1985), pp. 62–63.
13. Tyson (1992), p. 163.
14. Klepper (1990), p. 777.
15. McIntyre (1992), p. 36.
16. Klepper (1990), p. 778.
17. Tyson (1992), p. 167.
18. Newhouse (1982), chap. 6, 9.
19. Newhouse (1982), p. 31.

20. *Aviation Week and Space Technology,* March 17, 1997, p. 64.
21. *Financial Times,* February 18, 1994, pp. i, v.
22. Airbus Industrie (1997), pp. 4, 9.
23. Ibid., p. 5.
24. *Financial Times,* February 18, 1994, pp. i, v.
25. Airbus Industrie (1997), p. 13.
26. Ibid., p. 5.
27. Ibid., p. 1.
28. *Financial Times,* August 25, 1999.
29. *Asian Pulse,* August 19, 1999.
30. McIntyre (1992), chap. 7; and Tyson (1992), p. 172.
31. Tyson (1992), p. 169.
32. Mowery and Rosenberg (1989), pp. 178–186.
33. Tyson (1992), pp. 169–172.
34. *Los Angeles Times,* February 28, 1999, pp. C1, C13.
35. Thornton (1995), pp. 81–91.
36. Despite numerous missteps, the decline and ultimate demise of McDonnell Douglas as an airframe supplier can be attributed to its failure to offer a broad product line. While both the smaller MD-80/MD-90 and the wide-body MD-11 had their supporters, a huge gap in both range and seating capacity separated the two.
37. With the exception of the 717, formerly the MD-95, which Boeing acquired along with McDonnell Douglas.
38. Mowery and Rosenberg (1989), p. 176.
39. An Airbus print advertisement lampoons Boeing's smaller narrow-body fuselage with a headline pondering: "How would other aircraft manufacturers design your home?" and displays humorous images of a besuited passenger cramped by a diminutive door, a miniscule chair, and a Lilliputian bathtub. The text notes: "For although people are generally bigger these days, other aircraft still stick to the same puny dimensions as their 1960's predecessors," a thinly veiled swipe at Boeing.
40. *Aviation Week and Space Technology,* March 12, 1979.
41. *Aviation Week and Space Technology,* August 18, 1997.
42. *Aviation Week and Space Technology,* July 13, 1992.
43. *Aviation Week and Space Technology,* November 9, 1987.
44. Extended Twin Overwater Operations (ETOPS) certification (of a twin engine airliner's reliability over water) of the A330 was eased by its commonality with the in-service A340. *Aviation Week and Space Technology,* November 23, 1992.
45. The adoption of cross-crew qualification by other airlines has been hindered by labor agreements where pilot pay is frequently based upon size and weight of the aircraft.
46. *Aviation Week and Space Technology,* June 16, 1997.

47. *Aviation Week and Space Technology,* August 12, 1996.

48. *Aviation Week and Space Technology,* June 27, 1994.

49. The Federal Aviation Administration (FAA) has approved cross-crew qualification for the A320 and A330/A340. *Aviation Week and Space Technology,* August 29, 1994. See also *Aviation Week and Space Technology,* July 13, 1992.

50. *Aviation Week and Space Technology,* January 14, 1991.

51. *Aviation Week and Space Technology,* June 2, 1997.

52. *Aviation Week and Space Technology,* June 28, 1993.

53. *Aviation Week and Space Technology,* August 12, 1991.

54. *Aviation Week and Space Technology,* December 13, 1993.

55. *Aviation Week and Space Technology,* November 21, 1997.

56. *Aviation Week and Space Technology,* December 4, 1995.

57. *Aviation Week and Space Technology,* July 21,1997.

58. *Aviation Week and Space Technology,* June 16, 1997.

59. *Aviation Week and Space Technology,* April 20, 1992.

60. *Aviation Week and Space Technology,* July 7, 1994.

61. *Aviation Week and Space Technology,* June 28, 1999.

62. *Aviation Week and Space Technology,* April 11, 1994.

63. Other "Gang of Eight" carriers included American, British Airways, Delta, Qantas, and United. Sabbagh (1996).

64. *Aviation Week and Space Technology,* January 15, 1996.

65. *Aviation Week and Space Technology,* November 20, 1995.

66. *Aviation Week and Space Technology,* June 28, 1999.

67. *Aviation Week and Space Technology,* February 1, 1993.

68. *Aviation Week and Space Technology,* June 16, 1997.

69. *Aviation Week and Space Technology,* July 4, 1994.

70. *Aviation Week and Space Technology,* March 4, 1991.

71. *Aviation Week and Space Technology,* November 27, 1995.

72. *Aviation Week and Space Technology,* November 25, 1996.

73. *Financial Times,* February 25, 1992.

74. *Aviation Week and Space Technology,* March 27, 1992.

75. *Asian Pulse,* August 19, 1999.

76. *Seattle Times,* October 7, 1999.

77. *Aviation Week and Space Technology,* June 3, 1991.

78. *Aviation Week and Space Technology,* December 4, 1995.

79. *Aviation Week and Space Technology,* July 4, 1994.

80. *Aviation Week and Space Technology,* June 16, 1997.

81. *Aviation Week and Space Technology,* May 6, 1996.

82. *Aviation Week and Space Technology,* November 20, 1995.

83. *Aviation Week and Space Technology,* January 23, 1995.

84. *Aviation Week and Space Technology,* October 26, 1992.

85. McIntyre (1992), p. 141.

86. *Aviation Week and Space Technology,* May 12, 1986.

87. *Aviation Week and Space Technology,* March 6, 1989.
88. *Aviation Week and Space Technology,* September 2, 1996.
89. *Air Transport Intelligence,* October 6, 1999.
90. *Aviation Week and Space Technology,* June 21, 1976.
91. *Aviation Week and Space Technology,* January 14, 1991; May 13, 1991; December 24, 1990.
92. *Aviation Week and Space Technology,* November 15, 1993.
93. *Aviation Week and Space Technology,* June 15, 1987.
94. *Aviation Week and Space Technology,* March 10, 1986.
95. *Air Transport Intelligence,* September 13, 1999.
96. *Aviation Week and Space Technology,* July 17, 1978.
97. *Aviation Week and Space Technology,* April 16, 1979.
98. *Aviation Week and Space Technology,* July 4, 1994.
99. *Aviation Week and Space Technology,* November 4, 1991.
100. *Aviation Week and Space Technology,* April 29, 1996.
101. *Aviation Week and Space Technology,* June 20, 1994.
102. *Aviation Week and Space Technology,* March 17, 1997.
103. *Aviation Week and Space Technology,* November 25, 1996.
104. McIntyre (1992).
105. *Aviation Week and Space Technology,* October 26, 1987.
106. *Aviation Week and Space Technology,* July 4, 1994.
107. *Los Angeles Times,* February 28, 1999.
108. *Aviation Week and Space Technology,* June 21, 1976.
109. *Aviation Week and Space Technology,* June 8, 1981.
110. *Financial Times,* July 28, 1999.
111. *Seattle Times,* August 4, 1999.
112. *Aviation Week and Space Technology,* June 29, 1987.
113. *Aviation Week and Space Technology,* March 12, 1979.
114. *Aviation Week and Space Technology,* August 18, 1980.
115. *Aviation Week and Space Technology,* January 9, 1995.
116. *Asian Pulse,* April 21, 1998.
117. *Aviation Week and Space Technology,* November 21, 1994.
118. *Aviation Week and Space Technology,* July 1, 1996.
119. *Aviation Week and Space Technology,* December 21, 1987; August 29, 1994; October 16, 1995.
120. *Aviation Week and Space Technology,* April 5, 1993; *Financial Times,* November 18–24, 1998.
121. *European Report,* December 20, 1995.
122. *Aviation Week and Space Technology,* January 15, 1996.
123. *Financial Times,* February 25, 1992.
124. *Financial Times,* January 8, 1993.
125. *Aviation Week and Space Technology,* October 28, 1996.
126. *Air Transport World,* October 1, 1998.

127. *Air Transport Intelligence,* October 8, 1999.
128. *Aviation Week and Space Technology,* May 3, 1976; April 5, 1982.
129. *Financial Times,* January 22, 1997.
130. *Air Transport World* (March 1998).
131. *Asian Aviation News,* September 20, 1996.
132. *Air Transport World* (March 1995).
133. *Aviation Week and Space Technology,* May 4, 1981.
134. *Market Reports,* May 14, 1993.
135. *Aviation Week and Space Technology,* October 11, 1993.
136. *Aviation Week and Space Technology,* March 10, 1997.
137. *Aviation Daily,* August 13, 1999.
138. *Aviation Week and Space Technology,* April 28, 1997.
139. *Aviation Week and Space Technology,* February 24, 1997.
140. Tyson (1992), p. 185, citing a Harvard Business School case study.
141. Conversely, Aérospatiale's exceedingly rosy forecast of $280 billion in A380 sales versus an $8 billion initial investment may be intended to soothe apprehensive partners. *Aviation Week and Space Technology,* September 9, 1996.
142. *Aviation Week and Space Technology,* April 28, 1997.
143. *Aviation Week and Space Technology,* March 15, 1976.
144. Including a twelve-year lease and landing rights in Korea, special trade agreements in India, and nuclear power plants in Iran. *Aviation Week and Space Technology,* April 3, 1978.
145. *Aviation Week and Space Technology,* August 7, 1978.
146. *Aviation Week and Space Technology,* October 23, 1978.
147. When the Thai government considered canceling an order of two A300s in favor of 767s, Airbus salesmen were alleged to have threatened to retaliate by slashing European import quotas for tapioca, its second most important export. McIntyre (1992), p. xix. Also see *Aviation Week and Space Technology,* July 25, 1977.
148. Eastern President Frank Borman addressed his staff following the deal: "If you don't kiss the French flag every time you see it, at least salute it. The export financing on our Airbus deal subsidized this airline by more than $100 million." McIntyre (1992), p. 46. On the sectoral agreement, see McGuire (1997), pp. 74–75.
149. Tyson (1992), p. 199.
150. Canada, Sweden, and Japan, all characterized by significant government involvement in the aviation industry, also backed the European position. See McGuire (1997), p. 78.
151. Tyson (1992), p. 201.
152. Thornton (1995), pp. 138–139.
153. Ibid., p. 141.
154. Thornton (1995), p. 136.
155. Ibid., pp. 136–137.

156. Tyson (1992), p. 207.
157. Ibid., p. 205.
158. Thornton (1995), p. 146.
159. *European Report,* June 5, 1993.
160. Japan did pressure TDA to purchase the A300B2 to ease the trade imbalance with Europe. *Aviation Week,* December 3, 1979. See *European Report,* May 1, 1996, and *Financial Times,* September 14, 1994.
161. *European Report,* July 31, 1994.
162. Thornton (1995), pp. 148–149.
163. *Aviation Week and Space Technology* (August 1993), and *Aviation Week and Space Technology,* February 28, 1994.
164. *Aviation Week and Space Technology,* August 30, 1993.
165. *European Report,* May 21, 1997.
166. *Aviation Week and Space Technology,* November 22, 1993; May 18, 1992; September 28, 1992.
167. *European Report,* December 22, 1993.
168. *European Report,* November 9, 1996.
169. *Flight International,* January 25, 1995.
170. *Flight International,* August 5, 1992, and *Aviation Week and Space Technology,* February 5, 1996.
171. *Financial Times,* September 1, 1994; September 27, 1994.
172. *Aviation Week and Space Technology,* November 20, 1995; November 27, 1995.
173. *Tacoma News Tribune,* August 11, 1999.
174. See chapter one by Aggarwal in this volume.

References

Airbus Industrie (1997). *Global Market Forecast, 1997–2016: Confirming Very Large Demand.* Blagnac, France: Airbus Industrie.

———(1998). *Airbus News from Asian Aerospace '98, Airbus Industrie.* Blagnac, France: Airbus Industrie.

———(1999). *Global Market Forecast.* Blagnac, France: Airbus Industries.

D'Aveni, Richard (1994). *Hypercompetition: Managing the Dynamics of Strategic Maneuvering.* New York: The Free Press.

Hayward, Keith (1994). *The World Aerospace Industry: Collaboration and Competition.* London: Duckworth.

Klepper, G. (1990). "Entry into the Market for Large Transport Aircraft." *European Economic Review* 34. pp. 775–803.

McGuire, Steven. (1997). *Airbus Industrie: Conflict and Cooperation in U.S.-EC Trade Relations.* New York: St. Martin's Press.

McIntyre, Ian. (1992). *Dogfight: The Transatlantic Battle Over Airbus.* Westport, CT: Praeger.

Mowery, David C. and Nathan Rosenberg (1989). *Technology and The Pursuit of Economic Growth*. Cambridge: Cambridge University Press.

Newhouse, John (1982). *The Sporty Game*. New York: Alfred A. Knopf.

Sabbagh, Karl (1996). *21st Century Jet: The Making and Marketing of the Boeing 777*. New York: Scribner.

Thornton, David Weldon (1995). *Airbus Industrie: The Politics of an International Industrial Collaboration*. New York: St. Martin's Press.

Tyson, Laura D'Andrea (1992). *Who's Bashing Whom? Trade Conflict in High-Technology Industries*. Washington, D. C.: Institute of International Economics.

U.S. Civil Aviation Manufacturing Industry Panel (1985). *The Competitive Status of the U.S. Civil Aviation Manufacturing Industry*, Washington, D. C.: National Academy Press.

Wright, A. J. (1984). *Airbus*. Shepperton: Ian Allan Ltd.

Chapter 8

Penetrating the Regulatory Thicket in Asia: Nonmarket Strategies in Banking and Insurance

Klaus Wallner

I. Introduction

A chief source of growth in the world economy is the ongoing reduction of barriers to the international movement of capital. Many Asian countries have greatly benefited from this development. However, those sustained high growth rates came to an abrupt end in 1997 with the onset of the Asian financial crisis. Yet, despite extensive public and private borrowing from foreign sources, as well as the particularly pronounced dependence on foreign markets for their exports, the market presence of foreign financial institutions in many of these countries is marginal at best. A study of the financial sector in Japan, South Korea, and Taiwan concludes:

> Domestic opening to foreign institutions has lagged far behind the liberalization of financial flows. Whether by policy design, structural conditions, or costs of entry, foreign firms have small shares of the domestic loan, deposit, and other financial markets in all three countries.[1]

This limited penetration by foreign firms into these Asian financial markets is one of the main reasons behind the push for multilateral financial market liberalization efforts. These efforts culminated in the World Trade Organization (WTO) agreement of January 1998 to open 95 percent of the international financial services market by March 1999.

European financial firms pursue market and nonmarket strategies to break through structural and market obstacles and exploit profit opportunities in Asia. This chapter sheds light on the use of nonmarket strategies in international market penetration by focusing on two cases: bank lending in Southeast Asia, and casualty insurance in Japan. Both cases come from industries with high barriers to entry, both formal and informal. In the first case, bank lending in Southeast Asia, European firms focused their market penetration efforts on nonmarket strategies and succeeded in gaining a significant market presence. On the other hand, in the Japanese insurance industry, European firms were unsuccessful in using nonmarket strategic approaches, either because they did not devote enough effort— focusing instead on market strategies—or because the nonmarket strategies available to them were inefficient.

The main reason for the case study approach lies in the heterogeneity of Asian countries and of industries in the financial sector. Asian countries differ widely in terms of their cultural, social, and business environments. A comparison of China and Hong Kong illustrates the magnitude of such differences. China has 1.2 billion citizens living in almost ten million square kilometers (3.9 million square miles), compared with Hong Kong's city state of 6.5 million people in one thousand square kilometers (390 square miles); China's population density is 132 people per square kilometer compared with Hong Kong's 6,568. Per capita GNP in 1997 was US$710 in China, compared to US$25,100 in Hong Kong.[2] China, despite its ongoing gradual reform, is still a relatively inaccessible market, while Hong Kong (along with Singapore) takes pride in being among the most liberal market economies in the world. Market outcomes reflect different social values everywhere in the world, but the stark differences between neighboring countries in Asia must be considered extreme. In China, 0.4 percent of the adult population has life insurance policies, while 100 percent of South Koreans do. In the face of such striking differences, it is uninformative to examine market penetration strategies of foreign firms and measure the outcome in terms of market presence at the regional level.

Furthermore, the financial sector includes industries that differ in fundamental characteristics. Success in the securities business requires a very high level of skill, familiarity with and access to the global capital markets, and sufficient critical mass to take on large capital issues. Retail banking, by contrast, benefits most from familiarity with local market conditions and proximity to the customer. Crucial for success in the insurance business is an extensive sales network, and first mover advantages characterize the strategic environment at the retail level in this industry.

In recognition of significant country and industry differences, this chapter describes the environment and competitive approach of European firms in two particular situations: casualty insurance in Japan, and bank lending in Southeast Asia. The former illustrates that market strategies were unsuccessful in the face of large informal barriers to entry, while the latter highlights the successful employment of nonmarket strategies in overcoming entry barriers. The financial sector offers a good testing ground for the use of market and nonmarket strategies because the option of serving the foreign market from abroad—that is, by exporting or investing—does not exist. In order to be involved in retail banking or insurance, a foreign firm must establish a local presence. The alternative to entering the market directly is often to forego entirely the profit opportunities offered in the market. Being forced to enter, firms are then more aggressive in their pursuit of both market and nonmarket strategies.

Of particular interest is a comparison between European and U.S. financial firms. While European insurance companies failed to attain significant market share in Japan, their American counterparts have had moderate success in their lobbying efforts for more open markets. This success derives from the heavier political weight of the United States, while individual European governments are not sufficiently influential to promote their home firms effectively. In the case of private bank lending, U.S. firms have proven less ready to incur loan losses for the sake of getting a share of subsequent, profitable business, while European firms had to pay this price in exchange for market participation in lucrative business areas. Again, the explanation put forward in this chapter is that relative to their European counterparts, U.S. firms have used more effective nonmarket strategies to gain market access.

The chapter is organized as follows. Section II describes the positional analysis of firms entering Asian markets: the environment in which foreign financial firms in Asia operate, and its implication for their choice of market, nonmarket, and organizational strategies. Section III describes these firms' tactical and strategic analyses, and compares their short- and long-term approaches compared to those of their U.S. and Japanese rivals. Section IV is a case study of foreign firms' strategies in the Japanese casualty insurance market. Section V concludes with analysis of the overall patterns shown in the strategies taken by European financial services firms in Asia.

Table 8.1 Presence of Foreign Banks in Selected Asian Countries

Country	Foreign Banks/Representative Offices	Branches
South Korea	52	68
Japan	92 have branches	over 100
	120 have only representative offices	
Hong Kong	157 representative offices	—
China	500 have representative offices or branches; 5 foreign subsidiaries	134 have branches
Laos	—	6 (Thai banks)
Philippines	—	13
Singapore	161	—
Thailand	several dozen have only representative offices	21 have branches
Taiwan	44 representative offices and branches	—
Vietnam	43 banks have only representative offices	24 have branches

Sources: Far Eastern Economic Review (1998), *Yearbook Asia 1998; Japan Business* (1994), San Rafael: World Trade Press; Hu and Zhy (1998).

II. Positional Analysis

Firm and Market Position

The business focus of foreign banks and insurance companies in Asia is revealed by their main customer base, which is compatriot firms with business interests in the region. The presence of the banks is primarily the result of following their customers. In most cases there is no significant local customer base. The participation of transnational companies in direct local intermediation is rare. Banks are mainly engaged in interbank loans and financing of international trade and foreign exchange business, while insurance companies serve the foreign investors from their home country and insure international transactions. In both industries it is very difficult if not impossible to build up a profitable retail presence. Table 8.1 illustrates this point with a look at the market presence of foreign banks in selected Asian countries. In some cases formal barriers reserve the turf for domestic firms, while in others first mover advantages or customer preference for home firms create barriers that may be prohibitively costly to overcome.

The foreign market share in virtually every Asian financial market is very low. The international market segmentation is due to both formal restrictions on foreign ownership of domestic companies and behavioral

Table 8.2 Percentage of Total Assets Owned by Foreign Banks

Country	Foreign Banks' Assets (as % of total assets)
Japan	1.5
Indonesia	4
Taiwan	5
South Korea	5.5
Thailand	7
Brazil	9.5
Mexico	10
Malaysia	15
Chile	22
Argentina	22.5
United States	23

Source: The Economist (1997) and *BIS* (1997).

barriers to entry. The shares of foreign-owned banks' assets in Asian domestic markets are shown in table 8.2. For comparison, the table also shows data from American countries because they also experienced systemic crises in recent years, and responded to their difficulties by opening up, to one extent or another, their markets to foreign participants.

Internationally segmented financial markets were the inherited market structure at the end of World War II. Throughout the 1950s and 1960s, for instance, German banks still channeled all financing of international trade through foreign correspondence banks rather than affiliated or subsidiary institutions. On the other hand, French and British banks benefited from strong colonial links to individual overseas countries and established local market presence earlier. U.S. firms did not have imperialist roots in foreign markets, and were therefore reduced to largely domestic activities in a situation of internationally segmented markets. Germany's and Japan's roles as culprits and losers of the war further hindered their establishment of an international presence. Japanese firms in particular saw their regional links severely damaged by wartime events. The Japanese government, following the U.S. approach after the war, implemented regulations that kept commercial banking, investment banking, and insurance strictly separate. A general theme in the postwar development of financial industries was that tighter regulation created market segmentation. This segmentation caused inefficiencies that necessitated protection from foreign entry. Thus, domestic social objectives were the root cause for not only segmentation of national markets, but indirectly also for international segmentation. However, the local inefficiencies have also worked to the advantage of more efficient potential foreign entrants. The entry situation pits the advantages of being a significant home player—such as consumer biases, connections to the regulator, and familiarity with local conditions—and its disadvantages, which stem from the lack of competitive incentives, against the foreign player's market position and barriers to entry. The combination of the latter two determines to what extent foreign market access requires nonmarket strategies. Apart from the foreign player's market position, all elements are history dependent, as the description of international market segmentation suggests.

The choice of the location of the regional headquarters, as well as the geographic composition of business, also reflects historical factors. Due to considerations of political stability and cost factors, most continental European and U.S. institutions located their regional headquarters in Singapore in preference over Hong Kong, while many UK firms still have strong historic customer interests in Hong Kong, and retain their local center there.

Due to their limited involvement in local Asian retail markets, European financial institutions have avoided the most severe impact of the Asian crisis, which resulted from the rising number of bad loans and the drop in new business. The fact that they are mainly serving nonfinancial firms from their home countries lends their activities a stable, long-term orientation. However, many foreign financials have built large assets in Hong Kong and Singapore to gain local footholds in Asia. Intended as a prudent way of testing the water in the more developed markets, many have overdone it in "embracing the tiger" and were hurt in the process. Nonetheless, the long-term interests of serving compatriot foreign investors in Asia, as well as the option value derived from current market presence for future expansion, in most cases outweigh the short-run negative impact of the crisis. While the drop in expected growth across the region has dampened the business outlook for most participants in the Asian markets, it has done so less for foreign financial institutions.

On the other hand, the crisis itself created new opportunities for foreign companies. One example of a windfall opportunity for foreign firms created by the crisis is the failing of many Japanese banks or security houses. Because of the Japanese system of lifetime employment in large firms, before the layoffs happened there was virtually no market for mid-career professionals seeking new employment. Foreign firms can seize this opportunity to increase their supply of skilled staff by hiring the former employees of bankrupt local firms. Also, foreign firms are eager to acquire long-sought immediate market presence by taking over a troubled local competitor.

Nonmarket Position and the Effects of Entry Barriers

Most Asia-Pacific countries have in place binding upper limits on foreign ownership of domestic companies as well as a variety of other impediments to entry. For instance, in South Korea foreign banks are not allowed to open legally independent subsidiaries, but can only operate branches and representative offices. The insurance industry, split into life and non-life sectors, has only been formally opened to foreign participation since 1988, and few foreign firms have been able to establish a lasting market presence. The government has made it difficult for foreign security companies to gain permission to enter the market, and has limited the variety of products they can offer. In Japan throughout the 1980s, foreign investors were legally barred from acquiring domestic banks. This legal constraint on direct investment forced foreign entrants to choose greenfield investments as a mode of entry. Unfamiliarity with the local market conditions, as well as the sunk costs involved in building a brand name, increased the startup costs of entering the market via a greenfield investment rather than by taking over a healthy established firm. Furthermore, under the pretense of protecting local consumers, domestic regulation in Japan prevented foreign banks from getting involved in the market for current account deposits. This restriction in retail activity in turn denied them access to low cost sources of funds, and translated into a severe disadvantage in the loan market. The real factors behind formal barriers to foreign entry are a combination of successful lobbying efforts of domestic rivals seeking to protect their markets, as well as culturally based suspicions against foreign firms. It is important to bear in mind that economic theory suggests that such protectionism bears high costs in terms of inefficiencies in the domestic economy, mainly affecting domestic consumers in the form of higher prices.

Many of the formal entry barriers can be overcome by obtaining special permits from the appropriate bureaucratic source of authority. Perhaps more significant in practice are the informal barriers facing potential foreign entrants. Most notable among such collusive practices are business groupings (*chaebols* in South Korea and *keiretsu* in Japan) the links of which are cemented by cross-shareholdings. Members of such groups of affiliated firms generally give their business exclusively to each other, bypassing the marketplace. Prices and services in these long-standing business relationships are often determined through complex cross-deals, such as automobile companies selling cars in special deals to the employees of large insurance companies in exchange for promoting the firm's insurance policies to its customers. The main bank, at the core of a group of diverse industrial firms, receives the lion's share of the lucrative business of the group, often in preference over competitive outside opportunities. In exchange, the main bank is obligated to step in and help any member of the group in the case of financial distress. It has been an uphill struggle for foreign companies to break into market environments dominated by such exclusive, long-term relationships. In China, entry barriers to foreign banks are more formal. In most cases, Chinese regulatory authorities limit foreign business activities to foreign exchange and international trade activities, or to long-term lending for large projects; only rarely are licenses given for the opening of branches or the establishment of affiliated companies.

The fallout from the Asian foreign financial crisis may have positive implications for the reduction of these entry barriers. From a long-term perspective, the crisis has two beneficial aspects for foreign firms eager to penetrate local markets. First, both as a result of foreign pressure—in particular International Monetary Fund (IMF) conditionality—and the need to increase domestic efficiency, formal barriers to foreign investors have been reduced. Governments privatize banks not only to allow and to encourage majority ownership by foreign investors, but also to make regulations more transparent. It is not immediately clear why foreign participation should result in more restructuring and increased efficiency of local operations. One possible channel is that foreign firms may have harder budget constraints, because it is socially more expensive to bail out foreign-owned firms. In such a case, the domestic society loses the portion of the bailout that goes abroad. Implicit government bailout guarantees were likely to have played a strong role in excessive lending in many Asian countries. For instance, it is often claimed that the reckless expansion of the South Korean conglomerates was based on low borrowing rates, as lenders considered them too large and too strategically important for the government to allow their failure. Because they are subject to harder budget constraints and free from capitalist obligations endemic in Asia, foreign banks have stronger incentives to monitor and take an active role in the restructuring of loan recipients in times of difficulty.

Governments also open their markets because they are short of capital, and because foreign presence increases competitive pressures in domestic markets. After a hesitant beginning, the benefits of increased openness to foreign investors became clear as investment began to flow in to the region in record amounts.[3] The second positive long-term consequence of the crisis is the weakening of informal barriers to foreign entry. Such barriers in Asia result prominently from the collusive practices of corporate groups like the *keiretsu* and *chaebols* in Japan and South Korea. These practices are widely blamed for the crisis itself. Foreign market access rises as these practices are changed and these groups are dismantled. For instance, Japanese life insurers have begun to sell portions of their cross-shareholdings that provided the glue to the tight business relationship between the sectors.[4] Any loosening of these links tends to open up entry opportunities for foreign financial companies.

Foreign firms have also benefited directly from local consumers' loss of confidence in domestic financial firms. The crisis exposed the weakness of domestic Asian regulatory structures. In such a situation, consumers perceive foreign banks and insurers as more secure, due to more effective regulation in the foreign home markets. Foreign firms thus benefit from the superior regulation and financial soundness of their home markets and use them successfully as a marketing tool.

In sum, the same factors that tended to keep foreign competition out of the Asian financial sector are credited with causing the financial crisis, and therefore the crisis itself helped to increase pressure to bring down such barriers. After initial reluctance foreign firms, unlike portfolio investors, have adopted a long-term view and have taken up entry opportunities selectively.

III. Strategic and Tactical Analysis

Firms have a choice of market and nonmarket strategies in pursuing a presence in foreign markets. Foreign firms are often disadvantaged within these markets regarding price, quality, innovative products, new sales channels, and the like. They must compensate for these disadvantages by utilizing other strategies. These alternative strategic and tactical approaches can be categorized as market and nonmarket strategies, according to the approach outlined in chapter one of this volume by Vinod Aggarwal. Drawing from the two case studies, the following analysis highlights some of the common approaches taken in integrating organizational, market, and nonmarket strategies.

Organizational Strategy and Tactics

Firms have used joint ventures and strategic alliances to facilitate production and distribution. A fitting example is the strategic alliance between Swiss Bank Corporation and Long-Term Credit Bank of Japan. In 1997 both agreed to take on mutual capital stakes and to develop a common range of activities in the Japanese market. Since then the number of such strategic alliances, which mainly involve European banks, have increased substantially. Also in 1997, Nippon Life and the U.S. firm Putnam entered into an agreement according to which Putnam will manage part of the investment portfolio of the Japanese life insurance company.

Market Strategy and Tactics

A common first strategy is to try to penetrate the foreign market relying on the strength of one's products and financial resources. Since products are relatively homogeneous—as often dictated by regulation—and since in retail banking and insurance there are no suppliers of inputs or individual buyers with significant market power, of Michael Porter's five forces only rivalry and the threat of entry remain.[5] Given the many obstacles involved and the large expense of establishing a significant presence in a foreign market, a strategy of rivalry is only available to large, financially strong firms. Such companies are likely to possess political clout with their home government and with the local and national government in the host country. They are likely to use these political resources to buttress their market entry efforts. The entry of the American International Group (AIG) into the Japanese insurance sector presents an example. Benefiting from both its innovative product range and lobbying efforts on the part of the U.S. government, AIG used its financial strength to build over time a significant retail presence in certain market niches.

Another strategy used by many firms is to limit their objectives to serving the foreign direct investment and international trade related needs of compatriot manufacturing and primary sector companies. Medium- and small-sized firms from all but very few globally influential countries have little choice, since they have neither access to political leverage from the home government, nor the ability to influence the local government on their own. This describes the majority of foreign companies registered in Asian financial markets, as very few foreign financials have seriously attempted breaking into the local retail markets.

Nonmarket Strategy and Tactics

Nonmarket strategies mainly include legal efforts to enhance foreign market access, and lobbying activities directed at the home government or at the foreign government and their regulatory organs. For example, in December 1997 the international financier George Soros granted the Russian government an emergency finance loan. Soon after, Soros was involved in several successful high-stakes bidding consortiums in Russian government privatization auctions. In Asian markets, however, the close relationships between local governments and industries often prevent foreign players from gaining direct access to local legislative power.

Another example is the strategy of foreign firms to lobby their home governments to push for foreign market access. This is more effective if the weight of the home government on the world stage is large, and hence, U.S. firms have used this more than European firms. At present individual governments in Europe are too weak to be effective in using confrontational tactics or diplomatic tools in favor of their firms. In contrast, U.S. industries have had varying success at promoting market liberalization through bilateral trade agreements, as illustrated by the well-documented examples of market access agreements in cars, semiconductors, and insurance.[6]

Combination of Market and Nonmarket Strategies

In the case of foreign lending in Asia, European firms followed an approach that combines both market and nonmarket strategies. By increasing their loans to both public and private entities in Asia, European banks attempt to create goodwill. The expectation is that, in the future, they will receive preferential treatment in the form of business and market access concessions. Creating goodwill can be considered a market strategy, because the European banks have accepted the losses that accrue from these lenient lending habits as the price for profitable future business relationships. It is also a nonmarket strategy in that the goodwill created can translate into future preferential treatment by the local government. In contrast, U.S. banks do not need to engage in such an expensive strategy because of the influence they have through their access to the U.S. government. Because they possess alternative nonmarket strategies that guarantee access to future opportunities, U.S. banks can afford not to offer nonperforming loans.

IV. Case Studies

International Bank Lending in Asia

Data on the exposure of private banks to Asia reveal two crucial insights. First, the lending share between European and North American banks is

Table 8.3   International Bank Lending to Asia by Nationality of Reporting Bank

	Total Foreign Lending in Asia (in U.S. $bn)	Percentage by Nationality			
		Europe	North America	Japan	Other
Mid–1996	337.8	40.4	10.5	34.2	14.8
End–1996	367.0	42.2	11.0	32.3	14.5
Mid–1997	390.5	43.9	10.1	31.7	14.3
End–1997	381.0	47.1	9.7	30.1	13.1

Source: BIS (1998).

surprisingly unequal, as table 8.3 shows. The size of the European and North American economies, as well as the size of their banking sectors, are of roughly comparable magnitude. Yet the market share of international bank lending of European banks has grown to around five times that of North American banks. This is all the more striking when bearing in mind that the profitability of banks, measured in terms of the rate of return on capital, is much higher in North America.

Second, during the height of the Asian financial crisis, some European banks actually increased their lending to Asia, and European financial institutions as a group reduced total exposure by much less than Japanese and U.S. banks. Table 8.4 gives the total loans to Asian countries by private banks in select countries.

Table 8.4 shows the total private loans by donor and recipient country from mid–1996 through the end of 1998. Until the end of 1996, Southeast Asia grew at a high rate, and foreign commercial banks were eager to increase their volume of lending in the region. In early 1997, however, the first signs of trouble emerged. The volumes of loans show a striking difference among the responses of the lending countries. After the outbreak of the crisis in July 1997, European banks reduced their lending by much less than their U.S. counterparts. Commercial banks in the United States headed for the exit as loans sharply lost value. A comparison of outstanding loans in mid–1996 and end–1998 shows that the volume of U.S. bank lending fell in most countries to around 30–50 percent of the initial volume, while that of major European country banks actually increased.

The World Bank, referring to the Asian financial crisis, notes that domestic institutional weaknesses in Asian countries "were aggravated by undisciplined foreign lending, which led to too much money chasing bad investments."[7] European banks did not maintain or even increase their lending because they were generally more optimistic about the recovery chances of their loans than were their U.S. counterparts. Rather, they saw this lending as a vehicle for the penetration of markets that were otherwise largely closed to them, as the following quote suggests:

> Germany's Deutsche Bank AG . . . made risky, temporary loans across Asia so it could be first in line for corporate bond deals . . . Loans were often made for more than sound banking reasons. In South Korea, a loan to a large conglomerate, or *chaebol,* was synonymous to making a loan to the government . . . it was a Korea Inc. loan. Many of the banks made some risky loans at first in the hope of building a relationship that would yield more business later. A loan might lead to investment banking fees or a lucrative bond offering later.[8]

European banks did not extend loans based solely on their individual merits, but for the sake of facilitating access to the lucrative corporate bond offerings or investment banking fees. For example, loans made in South Korea to a large *chaebol* were tantamount to making a loan directly to the South Korean government, as the boundaries between private and public entities were unclear. The foreign lender to a *chaebol* could in return, for lenient loan practice, count on future business deals involving either the *chaebol* or some government agency. Since the rewards of this initial lending can come years down the line and are not openly linked to the loans, the ultimate success of this nonmarket strategy is difficult to evaluate. European lenders have certainly been successful in demonstrating their long-term commitment to the Asian markets; good friends stay when times get tough for their partners. Individual banks, though, paid a steep price in the short- and medium-term for this attempt to create long-term market opportunities. Many European financial institutions are likely to have underestimated the actual cost of using lending to gain market access. The regional economic crisis brutally undercut the financial stability of many loan recipients.

Casualty Insurance in Japan

The Japanese non-life insurance market is the second largest in the world, with ¥10 trillion ($82 billion) in premium income.[9] The main products sold in the industry are voluntary and mandatory automobile insurance (which together comprise around 60 percent of the total premium income), personal

Table 8.4 International Claims of Bank for International Settlements (BIS) Reporting Banks on Individual Asian Countries in U.S. $Millions

		France	Germany	Italy	Japan	Holland	UK	United States
China	End-1998	8,171	6,896	1,442	15,115	2,572	6,525	1,895
	Mid-1998	7,961	7,249	1,507	17,485	2,670	7,809	2,095
	End-1997	8,203	7,884	1,416	19,589	2,608	8,194	2,521
	Mid-1997	7,299	7,278	1,396	18,731	1,632	6,906	2,933
	End-1996	7,548	6,014	1,198	17,792	2,330	5,705	2,688
	Mid-1996	7,632	5,374	1,156	17,383	1,673	4,506	2,250
Hong Kong	End-1998	9,724	22,397	1,932	38,669	5,325	28,127	4,741
	Mid-1998	12,584	24,053	3,432	54,623	8,006	32,792	6,080
	End-1997	15,584	28,047	4,018	76,272	5,728	34,194	8,832
	Mid-1997	12,777	32,204	5,295	87,354	5,553	30,063	8,693
	End-1996	12,280	26,811	4,410	87,462	4,554	26,216	8,665
	Mid-1996	13,444	27,286	4,711	91,042	4,147	24,823	8,660
Indonesia	End-1998	3,874	5,638	142	16,402	3,319	3,814	3,537
	Mid-1998	4,009	5,876	152	19,030	3,445	3,967	3,226
	End-1997	4,773	6,174	184	22,018	2,975	4,492	4,893
	Mid-1997	4,787	5,610	187	23,153	2,823	4,332	4,591
	End-1996	4,463	5,508	104	22,035	2,458	3,834	5,279
	Mid-1996	3,652	4,843	97	21,622	2,466	3,260	3,551
South Korea	End-1998	7,425	8,250	601	16,925	2,544	5,551	6,291
	Mid-1998	7,913	8,400	850	18,934	2,564	5,634	7,409

(continues)

Table 8.4 *(continued)*

	France	Germany	Italy	Japan	Holland	UK	United States
End-1997	11,135	9,616	806	20,278	1,939	6,924	9,531
Mid-1997	10,789	10,794	1,369	23,732	1,736	6,064	9,961
End-1996	8,887	9,977	1,208	24,324	1,926	5,643	9,355
Mid-1996	6,994	8,529	1,024	22,512	1,651	4,140	9,582
Malaysia End-1998	2,335	4,618	86	6,623	930	2,040	858
Mid-1998	2,391	5,160	135	7,905	841	1,613	1,149
End-1997	2,883	7,197	222	8,551	903	2,014	1,787
Mid-1997	2,934	5,716	314	10,489	1,059	2,011	2,380
End-1996	2,641	3,857	167	8,210	730	1,417	2,337
Mid-1996	2,408	3,195	187	8,131	614	1,218	1,896
Philippines End-1998	1,698	2,304	85	2,324	2,363	1,844	2,657
Mid-1998	1,780	2,161	88	2,308	3,860	1,775	3,025
End-1997	2,165	2,999	97	2,624	2,546	1,607	3,225
Mid-1997	2,012	1,991	97	2,109	1,018	1,076	2,809
End-1996	1,873	1,820	56	1,558	929	1,173	3,902
Mid-1996	1,381	1,475	54	1,402	777	782	3,351
Singapore End-1998	7,494	30,934	2,641	29,474	5,971	21,873	2,337
Mid-1998	7,303	26,299	4,331	33,558	7,567	21,583	2,967
End-1997	13,316	39,177	4,882	58,649	5,445	24,943	3,786
Mid-1997	15,399	38,351	8,382	65,035	8,661	25,245	4,465
End-1996	11,771	40,767	7,139	58,809	6,679	22,523	5,727

(continues)

Table 8.4 (continued)

		France	Germany	Italy	Japan	Holland	UK	United States
	Mid-1996	12,684	36,609	6,092	58,784	8,769	21,011	6,040
Taiwan	End-1998	3,244	2,630	518	2,143	3,088	3,393	1,225
	Mid-1998	4,201	2,439	515	2,552	2,790	3,495	1,516
	End-1997	5,505	2,670	589	3,516	2,447	3,027	2,208
	Mid-1997	5,150	3,001	686	3,008	1,108	3,161	2,507
	End-1996	4,611	2,628	483	2,683	1,169	2,773	3,182
	Mid-1996	4,707	2,620	405	2,452	1,167	3,289	3,287
Thailand	End-1998	3,552	4,687	529	22,437	1,594	1,775	1,358
	Mid-1998	3,943	5,286	315	26,120	1,792	2,088	1,757
	End-1997	4,718	6,028	338	33,180	1,675	2,361	2,533
	Mid-1997	5,089	7,557	431	37,749	1,634	2,818	3,997
	End-1996	4,583	6,914	461	37,525	1,566	3,128	5,049
	Mid-1996	4,367	6,381	433	37,552	1,424	3,070	4,433
Vietnam	End-1998	516	218	0	251	205	215	95
	Mid-1998	449	242	2	213	212	218	80
	End-1997	500	296	2	257	214	135	123
	Mid-1997	373	316	0	241	139	130	101
	End-1996	390	309	1	249	136	99	184
	Mid-1996	339	269	0	249	137	123	131

Source: BIS (1997, 1998, 1999).

accident, and fire insurance policies. Large-scale claims from wartime destruction forced the Japanese government to consolidate the industry from its fragmented structure into only sixteen firms. Regulatory oversight over the industry lies in the hands of the Banking Bureau of the Ministry of Finance (MoF). The policy of MoF has been to preserve this structure by prohibiting new entries. The largest five companies have a combined market share of over 50 percent, and operate generally at lower costs than their smaller rivals do.

This protection results in high cartel rents. The average combined ratio of the Japanese "quoted sector"—that is, that of the fourteen largest companies listed at the Tokyo Stock Exchange—over the fiscal years 1984–1995 is 97.3. By contrast, the combined ratios of the United States and UK over the past twenty years have been 108.2 and 107.5 respectively.[10] Japan has consistently been the only major industrialized country where insurance underwriting is profitable. In other markets, profits are made not on the risk-coverage per se, but on the investment of the premium pool. This profitability is even more impressive because protection keeps products homogeneous and results in significant cost inefficiencies. In fact, the expense ratio—defined as expenses divided by total premia—of Tokio Marine, the most efficient firm, over the past years is about four points lower on average than that of the industry average. The MoF has consistently pursued a no dropout policy, setting rates at levels that keep even the least cost-efficient firm alive. As a result, no domestic company has failed since World War II. The low diffusion of insurance products in Japan is at least in part a result of the high premium rates.[11]

As of December 1998, there were twenty-nine foreign companies licensed in the market, with a combined market share of close to three percent. Their strength was concentrated in the so-called third area of sickness and nursing care insurance. This area contains sections of life as well as non-life insurance policies. Life insurance pays a regular amount, predetermined, in a certain contingency, while the amount paid out under a non-life policy in the claim case indemnifies the holder up to a certain amount for damages sustained. The third area policies are accident policies, but they result in regular payments that may be predetermined, such as home nursing coverage. These are relatively recent additions to the insurance landscape, and the innovative strengths and experiences in their more advanced home markets give foreign insurers an edge in this field over local competition.

The market is characterized by formal openness and, at the same time, high informal entry barriers. The *Insurance Business Law* provides equal access to the market for domestic and foreign firms. The main difference lies in the requirement of maintaining a minimum amount of capital inside the country, which is more onerous for foreign insurers given their small market share. There is a general requirement to get a license from MoF, and it has been a constant policy of MoF to license on average only one foreign company per year.

Although the rating associations formally have only advisory power, in practice they carry out crucial functions delegated to them by the MoF. Foreign insurers were not allowed to join the associations until 1994. Until then they received crucial information after their domestic rivals, and they were excluded from influencing industry-wide policies such as rate proposals. Even today the use of foreign actuarial material is not allowed in connection with a new product application. This regulation discriminates against foreign entrants with a market share that is too small to warrant incurring the large costs of supporting new product applications with local actuarial statistics.

Most large insurers in the concentrated Japanese market are affiliated with one of the major *keiretsu,* or business groupings. As part of the affiliation, the insurer holds significant stakes in other group members and is guaranteed the insurance business from the group. Foreign firms have long argued that this constitutes an entry barrier since they are effectively foreclosed from a large portion of the market.

Positional Analysis: Nonmarket Restrictions on Competition

Premium Rates.   The Fire and Marine Insurance Rating Association of Japan determines the uniform premium rates for fire, personal accident, and inland transit policies, subject to MoF approval. The Automobile Insurance Rating Association of Japan sets the industry-wide rates for automobile insurance policies, which account for about 60 percent of total premium income of the industry. In addition, the rates for marine, hull, cargoes, aviation, atomic energy, and engineering are not fixed by law, but agreed upon and applied uniformly by the industry. Formal, industry-wide consultations in organizations such as the Japanese Hull Insurers' Union, the Union of Machinery Insurers of Japan, and the Japan Atomic Energy Insurance Pool, as well as informal meetings, are sanctioned by the MoF and exempted from application of the Anti-Monopoly Law.[12] Thus, the rates of most standardized products are fixed.

New Products.   The rate-setting organizations function as fora for firms to coordinate the terms and conditions of all major insurance lines, with the result that all products in the industry are homogeneous. Furthermore, the

MoF only rarely approves new product types. No firm has an incentive to engage in costly innovation because the MoF refers all applications to the rate-setting associations for consideration, thus sharing among all firms strategic information about new product developments.

Distribution System.   Until 1996, insurance brokers were not allowed in Japan. Today the main distribution channels, in addition to the direct sales of group policies to affiliated companies, are coexisting exclusive and independent agents. In practice there is a perception among agents that if they promote a foreign entrant's product, the large domestic firms on which they depend for the main part of their income will withdraw their policies from the agency or otherwise express their disapproval.

Entry.   The MoF has generally pursued a no-entry policy. As in the banking industry, it has maintained a system in which all firms have been allowed to grow at similar rates. As a result, the market shares of the twenty-two domestic firms have remained fairly stable over time.

Trade Association.   The MoF informally delegates critical regulatory tasks to the trade associations. This practice includes the determination of premium rates and policy conditions, and the assessment of pending product applications. While the power of final approval rests with the MoF, its Insurance Division is understaffed, and the firms can often influence the rule-making process.

Regulation.   The power of the MoF derives from statutes that are worded in general form, the interpretation of which gives it wide-ranging regulatory authority over the industry. Most regulation consists in practice of a highly discretionary, situational set of unwritten ad hoc rules.[13] As long as the industry benefits from regulatory protection, the firms allow the MoF to utilize administrative guidance (*gyoosei shidoo*) to manage the cartel. By virtue of its informal and non-compulsory nature, there is no effective procedure for raising complaints in situations where a company feels subjected to unfair regulatory treatment. Administrative guidance often takes the form of a phone call from an MoF official to a middle-level manager at a firm, or is communicated in the course of one of the regular—almost daily—visits of company employees to the MoF. Such regulation is an effective tool to discipline deviant cartel members, as there is no transparent evidence and hence no recourse to ordinary grievance procedures.

Nonprice Competition. Regulation protects cartel rents by preventing nonprice competition. For example, only in 1996 did the government begin to allow partial comparative advertising. Even so this advertising, as well as other forms, has been mainly coordinated industry-wide.[14]

Regulatory Protection as Part of an Implicit Contract[15]

The fact that the MoF exercises the combined functions of financial sector oversight and the drafting of the national budget gives it significant political leverage. This autonomy must be borne in mind when analyzing this protective regulation. Essentially, the MoF is involved in a collusive deal with the domestic insurance industry. This partnership allows domestic firms to survive despite their inefficiencies, and gives them freedom from investigative and corrective actions by political bodies.

MoF protection consistently delivers substantial rents for the industry. There are several channels through which the firms pay for this protection. There is limited but increasing evidence suggesting that direct side-payments occur.[16] Also, there exists the practice of hiring retired insurance regulators (*amakudari*). The MoF enjoys remarkably harmonious interactions with the industry. Valuable payment comes in the form of the firms' submitting to costly directives that serve the wider interests of the financial regulator. The informal and discretionary nature of regulation allows the MoF to leverage its powers into areas where it is not given formal authority.[17] As long as no embarrassing failures occur, the MoF can exercise its powers freely. However, if something goes visibly wrong, for instance if one of the companies under the supervision of the Ministry should fail, politicians could assert formal authority over the MoF to satisfy their constituent voters.[18] Therefore, informal MoF power can compel the industry to excessive safety. Excessive safety levels are verifiable, as measured by the solvency margin standard—a risk-weighted assessment of the capital adequacy of each insurer. The industry average solvency margin standard has been estimated as 432.6, which indicates an overcapitalization by international safety standards of about 400 percent.[19]

It is important to bear in mind the dynamic nature of the interaction between the MoF and the industry. If the players interacted only once, the outcome would be drastically different. Given any level of regulatory protection, the industry would maximize its profits without cooperating with the MoF. On the other hand, given any level of cooperation, the regulator would be best off not granting any protection. The structure of this one-shot game is that of a static Prisoner's Dilemma. Only repeated interaction allows the sustaining of an implicit contract in a stable fashion over time. Since there is a static noncooperative equilibrium with zero rents for both industry and firm, there exists a credible threat of defection to this equilibrium in response to observed breeches of the implicit contract. For instance, in this way the MoF can oblige profitable members of the industry to bail out an underperforming rival. Should the firms refuse to incur this cost, they would face the consequence of losing the rents from regulatory protection in the future.

By submitting to a largely discretionary type of regulation, the firms allow the MoF to tailor punishments and rewards in a firm-specific manner. This gives the MoF an effective means to manage the cartel, and allows for a more profitable implicit contract on both sides. Owing to the extralegal character of regulation via administrative guidance, firms have no real complaint procedure, since compliance is voluntary, and the sanctions enforcing it are the future obstruction of business. Administrative guidance is in practice an extremely powerful tool to enforce collusive discipline. Only by granting MoF such power can the firms then retain their abilities to exploit the profit potential of the market.

Current Regulatory Changes and Implications

With the end of the bubble economy in Japan, and under the impact of the Asian financial crisis, the banking and insurance sectors are slowly being opened up to foreign competition.[20] U.S. insurers have successfully exploited the weakness of the financial sector since late 1996 to lobby forcefully for market opening measures. While European firms have echoed their demands, it is the U.S. negotiating effort that has yielded significant concessions. The harmful consequences of close relationships between regulators and firms in Japan have long been attacked by U.S. trade negotiators as market entry barriers.[21] Evidence on the nonmarket strategic efforts of U.S. insurance companies comes from the United States Trade Representative report which concludes that "the government of Japan still engages in an excessive level of regulatory activity in many sectors, which has a detrimental impact on market access."[22] In November 1996 Prime Minister Hashimoto announced a "Japanese Big Bang," which gave the ongoing trade negotiations on insurance a shot in the arm. The resulting bilateral insurance agreement of December 1996 aims at complete financial deregulation by 2001—in contrast to the outcome of prior U.S. government efforts to increase market access for U.S. insurance companies in the Japanese market. In particular, the 1994 agreement has been judged a complete failure, and a revision of the Insurance Business Law in 1996 has also not produced any changes in the competitive environment.[23] The reason is that both leave the centerpieces of the regulatory protection in place: fixed premium rates, and the discretionary power the regulator wields over the insurance industry.

The key measures in the 1996 agreement eliminate the power of rating associations to fix rates and curtail direct MoF intervention in the business of regulated industries. That the expected impact is severe is underscored by the heavy losses suffered by Japanese casualty insurance stocks upon the announcement of the insurance agreement. The measures were fiercely protested by the insurance industry. *The Economist* commented on these protests with customary satire: "There are few things so pleasing to the ear as the wail of a dying cartel."[24] The following are the most significant changes in the regulatory environment.

Price Liberalization.   The rates in personal fire and automobile insurance, which account for almost two-thirds of the premium total, will no longer be determined by the rating associations. Although gradually implemented, these measures promise to increase the ability of insurance firms to engage in price-based competition.

A Unified Insurance Sector.   Since April 1997, life and non-life insurance companies are allowed to enter their traditionally separate fields of business via subsidiaries. Mutual entry in the so-called third area (sickness, injury, and nursing care lines, which lie in the gray area of demarcation between life and non-life) was delayed until 2001.

Product Innovation.   A limited range of predetermined new products will no longer require individual approval by the MoF before being marketed. The extent to which products fall in this category is determined by growth in years, which should eventually comprise most of the big product lines.

Investment Liberalization.   Investment regulations have restricted all firms, mainly due to the high degree of mandated liquidity. The liberalization measures in this area allow firms to take better advantage of international investment opportunities and the financial instruments presented through a liberalized global capital market. This freedom will generate divergence in the financial performances of insurers, and hence will produce greater competition among them due to consumer sensitivity to differing investment records. In this case, foreign firms have an advantage that results from their deeper familiarity with international deregulated capital markets.

Disbursement of Capital Gains. With the aging of Japan's general population, the savings function of insurance policies is gaining importance relative to the insurance component. The permission to use capital appreciation to pay dividends to policyholders increases flexibility in investment strategy, since companies can reduce their focus on interest and dividend income. As a result, companies can distinguish themselves from competitors via superior investment performance, and hence higher returns to policyholders.

Insurance Brokers and New Sales Channels. Brokers are independent, and advise their clients on the most suitable and competitive insurance product to meet their needs. The introduction of brokers tends to increase competition among insurers. Furthermore, independent brokers are less susceptible to companies' pressure to shun the product lines of their foreign rivals. The introduction of new sales channels will facilitate the entry of new firms as is the case, for example, in mail-order sales of insurance policies.

The implicit contract theory explains the phenomenon of regulatory protection as well as the eventual breakdown of such an understanding. Protection under the terms of the implicit contract proves increasingly costly in the post-bubble environment. Embarrassing corporate failures in the banking sector drew public attention to regulatory inefficiency and sub-optimal, inefficient performance. Monitoring by politicians has become more aggressive since political instability weakened the traditional links between the longstanding governing Liberal Democratic Party and the MoF. The regulator emerged significantly weakened from the bursting of the financial bubble that built up during the high-growth period. Essentially, with the public monitoring bureaucratic behavior more closely, the MoF has less protection to offer to the firms. The industry has found it no longer optimal to cooperate with MoF requests as much as they had, thus revising downward the terms of the implicit contract.

Recent trends in the Japanese insurance market highlight the success of nonmarket strategies of U.S. firms, and on the other hand the failure of the pure market strategies of European firms. The U.S. government, lobbying on behalf of its national insurance industry, won much deregulation, and U.S. firms are the only foreign insurers with significant and growing market share.

EU insurers, having no access to equally powerful nonmarket strategies, were restricted to market strategies, and did not succeed in making inroads into this lucrative market.

It is still too early to assess the response of European insurance companies to these regulatory changes. While it is clear that such fundamental deregulation makes the use of market strategies more attractive, certain restrictive patterns remain. In such a concentrated industry, the market leadership still rests with the large domestic firms. To be able to benefit fully from cross-entry opportunities (e.g., for a non-life insurer to enter the life insurance market), a firm must be large enough to be able to recuperate the entry costs and build up productive capacity and sales channels through operating profits. Clearly, innovative sale channels benefit foreign firms, as they abolish the need to invest in more expensive, traditional sales networks. An early indicator that European firms are expected to gain strength and win business from domestic rivals is the fact that the stock prices of Japanese insurance companies drastically fell upon the announcement of the deregulation measures and of the U.S.-Japan Insurance Agreement.

V. Conclusion

Some clear general lessons emerge from the two cases. In general, firms utilize the most profitable combination of market and nonmarket strategies in their pursuit of profit opportunities in foreign markets. The case studies examined bank lending in Southeast Asia and casualty insurance in Japan, suggesting that the relative use of these strategies depends more on the national origin of the foreign firm than on the sector of activity. The national origin largely determines the power of nonmarket strategies, such as lobbying on the part of the home government. If the domestic government plays a relatively weak role on the international stage, firms may be forced to rely on market strategies. This puts them at a disadvantage before rivals from stronger countries, such as the United States, that can compete in both dimensions.

In penetrating Asian markets, the main strengths of foreign competitors—new products and innovative sales methods—are often of limited use because they are blunted by obstructive regulation. For instance, in the non-life insurance business, differentiating car insurance policies according to driving records or age allowed more product variety to consumers who were unsatisfied by the limited choice of products offered by the protected domestic industry. However, this market strategy can only be applied if the market is liberalized sufficiently. In the Japanese insurance case, European firms failed to achieve significant market presence through the use of these market strategies, while U.S. firms effectively used government lobbying to obtain market-opening concessions.

The two cases emphasize that a crucial determinant of the choice between market and nonmarket strategies is national origin. Because they could not compete using the nonmarket strategy of government lobbying, European firms looked elsewhere for their strategies. As the banking example illustrates, there may be sector-specific nonmarket strategies. By accepting losses in lending to finance the rapid expansion of Southeast Asian industry, European banks attracted large amounts of complementary business. They established a significant market presence in a sector where relationships are valuable, and individual transactions are not isolated from follow-up or unrelated business. The successful establishment of a strong market presence, and the familiarity with the local market that results, are of immense strategic value. This may justify the initial expense of over-lending.

While successful in the banking example, this strategy was not available to foreign firms in the Japanese insurance case. In that sector, high entry barriers forced firms to make use of the nonmarket strategies available to them. The availability of such strategies and their relative efficiency are qualities that vary according to market position and national origin, and drive the choice of which to apply.

A particular issue arose in the insurance case study regarding the simultaneous pursuit of market and nonmarket strategies. Some foreign firms, most notably AIG of the United States, were successful in developing new products in the so-called third sector. This market niche includes products with characteristics bordering on both life and casualty insurance policies. Foreign companies have successfully argued that allowing free entry in this area before other areas of insurance are liberalized jeopardizes their existence. It was a significant concession of the Japanese negotiators in bilateral trade talks with the United States to hold off with the planned liberalization of the third area until other insurance markets have been sufficiently reformed to allow small foreign firms to survive. This incident constitutes a visible and successful use of nonmarket strategy by U.S. insurers, and it illustrates the complex interaction between the market and nonmarket strategies of foreign firms in pursuing opportunities in a profitable foreign financial market.

To what extent do the insights from the two specific cases also apply to other markets? The formal barriers are comparable in other cases. Ministries of finance, which oversee the financial sector in most Asian countries, gererally retain the discretion to license foreign firms for participation in a financial sector industry. For example, the Japanese MoF has used this tool to keep foreign banks from competing in the retail-banking sector. This has had the consequence of forcing firms in that sector to face higher financing costs than their Japanese rivals, who have access to a low-cost local deposit base. This restricts the foreign firms' ability to compete in the loan market. Thus, restrictions on the competitiveness in one industry or sector can have an effect on the ability of firms to compete effectively in others as well.

Informal barriers derive from a wide variety of sources, and can take different forms. The creativity of a firm or an industry is often at its peak when it comes to protecting its profit base. In South Korea, industry groupings called *chaebol* have had the same effect of excluding potential foreign entrants as have the *keiretsu* in Japan. In virtually all countries, some type of implicit contract exists between large industries and their regulators, which guarantee the cooperation of the overseeing agency in protecting the interests of the industry. One thing in common to all Asian countries compared to Western industrialized nations is their weakness in monitoring their governments and public agencies. The informal powers and discretionary behavior that result allow such implicit contracts to flourish. Under these circumstances, informal entry barriers have in many cases proved effective barricades against significant competition from foreign entrants, successfully reducing the menu of market and nonmarket strategies available to potential foreign competitors.

Notes

1. Patrick and Park (1994), p. 345.
2. World Bank database, http://devdata.worldbank.org.
3. In the first five months of 1998, European FDI in Asia across all industries amounted to around US$4 billion, including significant investments in the financial sector by Commerzbank (Korea Exchange Bank; $249 million) and ABN Amro (Bank of Asia; $180 million). This contrasts with a slack 1997, when the Asian economic crisis first began to bite. *Financial Times,* June 15, 1998.
4. *Financial Times,* July 8, 1998.
5. Porter, M. E. (1980).
6. U.S. Chamber of Commerce in Japan (1997), *Making Trade Talks Work: Lessons From Recent History,* contains an evaluative survey of all recent bilateral trade agreements between the United States and Japan.

7. *The New York Times,* January 28, 1998.
8. Ibid.
9. BZW Research (1994). Currency conversion is as of March 27, 2001.
10. The combined ratio is the sum of expenses and claims, divided by total premiums earned. BZW Research (1994).
11. *Sigma,* a publication of Swiss Reinsurance Company, quotes the 1995 USTR Report that in 1992 Japan ranked tenth in the world in terms of per capita premiums. Japan's total premia value is one-fourth that of the United States, with half the population of the United States, and only little more than Germany, but with roughly twice its population (Total premia: United States, ¥38 trillion; Japan, ¥10 trillion; Germany, ¥8 trillion).
12. The original reason for exemption from the *Anti-Monopoly Law* is that certain risks are too large to be assumed by a single insurer alone, necessitating co- and re-insurance. This is a common justification for a similar exemption in many countries.
13. While difficult to quantify, administrative practice is estimated to constitute in excess of two-thirds of all regulation in some industries, and the insurance industry is certainly one of the most tightly regulated industries in Japan. See Schaede (1995).
14. Only once a year, during November, the non-life insurance industry conducts extensive advertising campaigns, mainly in concerted fashion.
15. This section draws on Wallner (1998).
16. Scandals involving kickbacks paid to career bureaucrats rarely become public, and certainly data are not available. In 1995 two small financial institutions collapsed. It turned out that two senior MoF officials had repeatedly accepted personal favors from executives of the troubled institutions, ranging from private plane vacation trips to all expenses-paid golf trips and direct cash payments, in exchange for the MoF planning a public bailout of the two companies. The top-level bureaucrats were not dismissed, but their careers as bureaucrats certainly took a turn for the worse. More recent scandals include officials from the MoF inspection department that were found accepting "wining, dining and golf" entertainment in exchange of advance information on unannounced control visits by the regulator. This even resulted in a raid of MoF offices by Tokyo public prosecutors in January 1998. Much of the side-payments are channeled through the trade associations, probably to avoid excessive competition between the members of the industry.
17. Examples of the wide range of opportunities for industry cooperation abound. For instance, MoF has a strong interest in keeping stockmarket movements under control, since they directly affect the strength of banks' balance sheets. When prices at the Tokyo Stock Exchange fell close to levels considered dangerous for many banks in 1993, MoF used informal regulatory pressure to prevent insurers from selling part of their stockholdings, as it has since done repeatedly. See *The Nikkei Weekly,* September 6, 1993. For

the same purpose, insurers were barred from issuing convertible bonds between 1989 and 1994, since the capital markets in this period were particularly shaky due to the fallout of the end of the bubble economy. See *Nikkei,*
May 24, 1993, and *The Nikkei Weekly,* April 18, 1994.

18. As a result of the bad-loan crisis in the banking sector, for which the MoF
    is held partially responsible, during 1996 a reform proposal circulated in the
    government proposing a split-up of the MoF's jointly held functions of
    drafting the national budget and overseeing the financial sector. While this is
    unlikely to become law (at least in this extreme form), this proposal illustrates
    what is at stake for the regulator if it fails to prevent instability. *The Nikkei
    Weekly,* June 10, 1996.

19. Credit Suisse Japan Research (1994).

20. There have been several attempts at deregulation in the Japanese insurance
    industry. After over three years of deliberations and research, the Insurance
    Council produced a report entitled "A New Role Demanded of the Insur-
    ance Industry" on May 29, 1992. See *Japan Insurance News* (1992). The revi-
    sion of the *Insurance Business Law* (1939) in April 1996 follows closely the
    recommendations of the report. In parallel, trade negotiations between Japan
    and the United States culminated in two agreements in October 1994 and
    December 1996.

21. USTR (1993, 1994, 1995).

22. USTR (1995), p. 144.

23. *The Economist,* June 15, 1996.

24. *The Economist,* December 21, 1996.

References

Barclays de Zoete Wedd (BZW) Research (1994). *Japan: Non-Life Insurance,* Reference Annual.

Credit Suisse Japan Research (1994). *Japan's Insurance Industry: Adversity Continues,*
Tokyo: Credit Suisse.

Hu, Rongbin and Hianwei Zhy (1998). "Bankmarkt China im Umbruch." *Die
Bank* (March).

Insurance Council (1992). "A New Role Demanded of the Insurance Industry" in
Japan Insurance News (July–December).

Marine and Fire Insurance Association of Japan, *Non-Life Insurance in Japan,* Fact
Book, various issues (Tokyo).

Non-Life Insurance Institute of Japan (1993). "Miscellaneous Casualty Insurance in
Japan" in *Japanese Practices of Non-Life Insurance Series,* Vol. III, 4th ed., (Tokyo).

Schaede, Ulrike (1995). "The Old-Boy Network and Government-Business Relationships in Japan" in *Journal of Japanese Studies* 21 (2) pp. 293–317.

Statistics Bureau of the Management and Coordination Agency (1993). *Annual Survey on Household Expenses.*

United States Trade Representative (USTR) Report (1993, 1994, and 1995). Washington, D. C.: USTR.

Wallner, Klaus (1998). *Implicit Contracts Between Regulator and Industry: The Case of Japanese Casualty Insurance* Working Paper 134. (New York: Columbia Business School, Center on Japanese Economy and Business).

Part Four

Conclusion

Chapter 9

Lessons from European Firm Strategies in Asia

Vinod K. Aggarwal

I. Introduction

Asia has long enticed foreign firms. This region includes many of the world's fastest growing markets, and promises to be a dynamic and fiercely competitive arena for decades to come. The regional currency crises of 1997–1998 frustrated but failed to diminish the ardor of foreign firms for the region. Both before and after the crises, firms have attempted to devise trade and investment strategies that would give them a competitive advantage over their rivals.

The objective of this volume is to present a novel framework to understand the market and nonmarket strategies that have enabled European firms to win in Asia. From an empirical standpoint, the book explores the political and economic constraints facing firms wishing to enter the Asian region, as well as strategies to secure European governmental support. It also includes a set of case studies of the software, auto, commercial aircraft, banking, and insurance industries to provide a basis for comparing and contrasting how firms in these sectors have attempted to enhance their competitive positions. In many cases, the case study authors have provided valuable comparisons of European firm strategies with American or Japanese firms, thus giving insight into the impact of national origin on competitive success.[1] These sectoral analyses also provide deeper insight into how firms have attempted to build effective relations with governments in the region and their home countries, as well as with regional institutions in Asia and Europe. In doing so, the objective has been to identify the most successful strategies for meeting the unique challenges of Asian markets.

This chapter is organized as follows. Part Two begins with a focus on the context within which European firms have operated, concentrating on the economic characteristics of the Asian market and the relative performance of European firms. This section also examines the political institutional arrangements that firms face in both the host and home markets. Part Three describes the market and nonmarket opportunities and constraints that prevail in the five sectors covered in this volume. Part Four assesses the theoretical and empirical aspects of the strategies and tactics pursued by European firms in Asia. Part Five concludes with a discussion of lessons that emerge from the book's analysis and offers directions for future research.

II. The Political and Economic Context: Competing in Asia

What is the nature of the strategic challenges and opportunities in Asia? In Part Two of this book, Shujiro Urata, John Ravenhill, and Cédric Dupont focus on different facets of the political-economic context within which European firms operate in Asia. While Urata is concerned with describing the performance of European firms and the changing Asian market, Ravenhill and Dupont examine the political context of lobbying in Asia and the European Union (EU), respectively.

The Economic Context

Despite recent ups and downs, the Asian market has been a highly attractive market for investment and trade. As Shujiro Urata notes in chapter 2, much of East Asia has been characterized by phenomenal growth, beginning with Japan in the 1960s. In the 1970s the newly industrializing economies (NIEs), such as Hong Kong, South Korea, Singapore, and Taiwan, experienced rapid growth, followed by a second tier of countries such as Indonesia, Malaysia, and Thailand in the 1980s. This economic growth has been the subject of intensive scrutiny by political scientists and economists. Factors such as macroeconomic policies, education, coherent state policies, and an outward oriented focus are just a few of the factors that have been used to account for this growth. More recently, while some analysts were very critical of the East Asian model in view of the 1997–1998 financial crisis, the rapid recovery of most countries in the region has reinforced the earlier view that these countries have pursued fundamentally sound economic policies. Moreover, the combination of Asian firms facing financial problems and pressures from the IMF to liberalize have created significant opportunities for foreign firms.

In this context, European firms have been drawn to the Asian market. Yet as Urata observes from his empirical analysis, the activities of these firms have been considerably less important in the aggregate than that of American and Japanese firms. For example, in trade, although European exports have kept pace with the overall growth in world trade, in 1998 Asia accounted for less than 5 percent of European exports. By contrast, Europe accounts for just over 15 percent of Asian exports. With respect to foreign direct investment (FDI), although Europeans have increased their aggregate position in Asia, along with the United States and Japan, they have experienced relative declines in FDI flows as intra-Asian investment grew rapidly in the 1990s. More recently, European firms in the aggregate increased their investment, and European banks became leaders in lending to Asia, as Klaus Wallner describes in chapter 8.

What factors account for the generally lower participation of European firms in FDI as compared to their American and Japanese counterparts? Urata argues that both geographical and cultural distance appear to be relevant in explaining patterns of trade and investment. He also comments on a study by the EU and the United Nations Conference on Trade and Development (UNCTAD) that evaluated patterns of European investment. This study identifies several key factors to account for the relative lack of European FDI penetration: the regulatory regime for FDI in Asia, structural characteristics of Asian host countries, transaction costs and economic distance, support from the home government, investment strategies, and preoccupation with regional integration. Of these factors, Urata dismisses the first three, arguing instead that European firms have not had a well formulated strategy particularly as compared to the Japanese and observing that they have been excessively preoccupied with European integration. He also agrees that European government support has not been as forthcoming as in the Japanese case, but he finds this argument to be less compelling. In examining the latter, Urata considers various European efforts to promote economic ties with Asia, and suggests that training programs and information dissemination efforts have yielded some benefits. The most important government-to-government effort, however, has been the Asia-Europe Meeting (ASEM) process discussed in detail by John Ravenhill—which brings EU and East Asian officials together in a series of meetings every two years to discuss mutual concerns.

The Political Context: Lobbying in Asia and the EU

Although the economic context and market strategies are undoubtedly critical in explaining the success of European firms in Asia, understanding both the host and home environments is essential for the development of nonmarket strategies. John Ravenhill discusses the first of these in chapter 3, focusing on the impact of regional arrangements within the Asian context. The second is addressed by Cédric Dupont in his examination of the political environment in Europe. The strategic approach he presents in chapter 4 provides firms an important roadmap for lobbying in the European Union.

Regional Arrangements in the Asian Context.   Regional institutions in Asia are much less developed than those of Europe. In focusing on the Association of Southeast Asian Nations (ASEAN), Asia-Pacific Economic Cooperation (APEC), and ASEM, Ravenhill attempts to gauge the extent to which firms should focus on these accords in their lobbying efforts, as opposed to a single-minded concentration on national governments. Beginning with ASEAN, Ravenhill traces the development of these arrangements. He argues that despite the creation of an ASEAN Free Trade Area (AFTA) in 1992, this accord's institutionalization process remains embryonic. The Secretariat is weak, and the only significant avenue of influence for firms appears to be the ASEAN Chambers of Commerce and Industry. Thus, argues Ravenhill, nonmarket strategies in the ASEAN countries are best focused on national member governments rather than the institution itself.

Turning to APEC, Ravenhill notes the close similarity between ASEAN's and APEC's institutional and decision-making structures. He argues that APEC has failed to develop significant independent decision making capabilities, relying instead on relatively ineffective peer pressure and socialization.[2] Finally, he notes that ASEM was developed in response to concerns that European firms were not benefiting from Asian markets as compared to their developed country counterparts. For the most part, East Asian governments saw this as an opportunity to counteract U.S. leverage in Asia. Yet ASEM has had only three official meetings, and business has shown little interest in its dealings. Concluding, Ravenhill suggests that while these institutions may have future promise as a locus of power, in their present state they have little impact on nonmarket strategies. Put

simply, firms may wish to focus their limited resources at the level of national governments, at least for the near future.

Lobbying in the EU.   Regional arrangements in Europe are in sharp contrast to those of Asia. As Cédric Dupont notes, lobbying in the European
context is not a simple matter. The complexity of EU institutions poses a
challenge to firms that wish to pursue nonmarket strategies. His chapter
provides a decision map for firms wishing to negotiate this maze to secure
EU support for their activities.

  After tracing the evolution of various institutions in Europe, Dupont
considers several key elements that influence the choices of firms. First,
firms must ascertain whether the Council of Ministers or the Commission
share authority in a particular issue area or whether they deal separately
with issues. In simple terms, the more developed the legal framework in an
issue area, the more important the role of the Commission. Second, firm
strategy is based on their aggregate resources; those who are relatively weak
are unable to lobby their targets directly. Third, firms with good national
connections have the capacity to focus on the Council, or within the
Commission can focus on the Commissioner's cabinets. Finally, sectoral
cohesion is a key variable: if it exists, even larger firms are tempted to lobby
with their counterparts, rather than going it alone. The combination of
these factors generates an enumeration of sixteen lobbying strategies and a
clear analytical guideline for firms. Dupont also provides a systematic empirical examination of several sectors to demonstrate how firms might in
practice go about deciding where to lobby.

III. Positional Analysis

Before firms can formulate a winning strategy, they must consider not only
the broader political-economic context discussed in Part Two, but also the
contours of the market in which they operate, their specific core competencies, and the nonmarket factors that affect their business. With respect
to each of these three elements, firms must take into account the nature of
their activities at the national, regional, or global levels. On this latter score,
I suggest in chapter 1 that firms must make decisions about locating their
trade or investment operations at the national, regional, and/or global level
and also decide on the target market for sales.

  To examine the opportunities and threats firms face at these three levels, I suggest in chapter 1 that a good approach to the examination of the

nature of markets is Michael Porter's "five forces model."[3] Using this model, the case studies consider the barriers to entry presented by firm rivalry, the potential of new competitors entering the market, threats presented by possible market substitutes, and the bargaining power of suppliers and buyers. The work on firm competencies is also relevant, but considerable debate continues over how one might best examine a firm's capabilities. While this question is less central to the interests of this volume, Gary Hamel and C. K. Prahalad's focus on "core competencies," which entail both tangible and nontangible capabilities, provides a useful entrée into understanding the abilities of firms.[4] Finally, with respect to nonmarket analysis, I consider David Baron's recent work that provides insight into the nonmarket environment of firms.[5] Baron argues that firms must be attentive to possible threats and opportunities arising from the nonmarket environment. Specifically, they must understand the issues raised, the interests of major groups, the institutional setting for policy resolution, and the information available to actors. Because these three sets of factors interact, firms attempting to succeed in the Asian markets must analyze systematically their market, core competencies, and the nonmarket environment in formulating and implementing strategy.

The case studies provide valuable positional analyses of several key sectors in which European firms have competed in Asia. Trevor Nakagawa observes that with the global proliferation of the personal computer and client-server architecture, the market dynamics in the software industry changed dramatically, leaving European software producers behind. Whereas they had succeeded in meeting the specialized needs of local clients with proprietary hardware designs, the new global software market has been marked by extremely rapid growth, increasing returns, open standards, network externalities, and pressure for compatible standards. In this international context, European firms' core competencies were not suited to the packaged mass-market software, and U.S. firms, with both cost and quality advantages, rapidly acquired dominant market shares in both Europe and Asia. However, many European firms found ways to leverage their local expertise by specializing in the lucrative service and systems integration business. However, these firms largely opted out of producing software, preferring to rely on the pervasive U.S. products.

Both Asian host and EU home governments provided important nonmarket contexts for software development in Asia. On the one hand, Asian countries have attempted to promote their own software production through provision of venture capital, the creation of "Software Parks," the organization of standards consortiums, attempts to cope with software piracy, and liberalization efforts to woo foreign firms. Meanwhile, European governments and the EU have created a variety of long-term IT consortia to promote the software industry and bolster the competitive abilities of European firms. Yet as Nakagawa notes, the relatively modest success of these projects seems to be explained by a tendency to focus on large established companies that often lack an export orientation. In short, the combination of these positional factors has left European firms in a relatively difficult position, with the notable exception of SAP.

The Asian auto market has proven highly attractive to foreign firms. As Nick Bizouras and Beverly Crawford note, rapid economic growth in developing Asia, and particularly in China, provide a contrast to the saturated markets in developed countries.

With respect to the market, local Asian suppliers are relatively weak, and consumers are in a relatively poor bargaining position in view of the high demand for autos. There is little threat of substitutes because local public transportation and trains cannot meet the demand for travel in rapidly growing economies. In terms of competition, Japanese firms have dominated most Asian markets, with the exception of China, thus making it the last open frontier. The auto market faces tremendous pressure for consolidation, with transnational mergers such as DaimlerChrysler and other cross holdings increasingly becoming the norm. Moreover, as competition increases, firms have been tempted to subcontract and source globally to reduce costs. Yet this market strategy works against success in China. Their research suggests that local allies are essential in the Chinese market because of an active, interventionist Chinese state that promotes technology transfer and the interests of its own local suppliers. This factor is particularly important in light of the devolution of power in China from the center to the regions. In response, firms must invest in nonmarket capabilities on a broad basis, including ties to local and regional government officials, if they wish to succeed in gaining market access. Those companies that have already developed core competencies in negotiating with authoritarian governments (such as Volkswagen in Eastern Europe) have a competitive advantage in China.

In their case study of Airbus, William Love and Wayne Sandholtz highlight the overriding importance of positional analysis in an industry characterized by high risk, enormous capital investment, and international competition. For Airbus, the Asia-Pacific market is attractive for its rapid growth and for the relative equality of its playing field. The absence of indigenous aircraft manufacturers leaves Boeing as Airbus's only competitor. Because no regional market alone is large enough to provide the large economies of scale necessary for a profitable aircraft industry, Airbus and

Boeing compete globally. Failure in the key Asian market could thus have been fatal for the more recently arrived Airbus. Another important element of the market is the high cost and the technology-intensive character of aircraft development and production. Aircraft producers must invest heavily in design, modeling, systems integration, and testing, and must carefully shape their aircraft to meet the specific needs of clients. At the same time, since each airplane model has an extremely long lifetime of production and sales, there is intense pressure to make designs flexible, so that an existing model can be stretched or shortened through modifications in the fuselage section. This is particularly important in Asian markets, because Asian carriers in the coming decades are expected to demand increasingly larger aircraft with greater seating capacities at a faster rate than U.S. or European airlines. Regarding nonmarket factors, the support of home governments has proved extremely important to both Airbus and Boeing. Neither company made a profit over its first two decades of business, but relied instead on government subsidies, grants, export financing, military contracts, tax exemptions, and other tools.

Klaus Wallner describes the financial services industry as facing significant formal and informal barriers in Asian markets. As a result, the foreign market share of Asian financial services markets is very small, with most foreign firms serving compatriot firms in the region. In the casualty industry in Japan, for example, foreign firms had only 3 percent of the market at the beginning of 1999, a figure that had not changed significantly in the 1990s. Asian markets tend to be highly segmented as a result of a history of tight regulation and protection of local, state-connected firms and banks. The economic crisis, while slowing down investment and business in the short-term, has had positive effects for foreign financial services firms in Asian markets. The restricted nature of their involvement kept them largely out of harm's way, and the failure or downsizing of many local banks or security houses presented an opportunity to hire locally connected and skilled employees. Most importantly, the fallout from the crisis generated pressures regionwide for liberalization of financial services, including for the reform of informal banking and corporate networks in South Korea and Japan, and the lowering of formal entry barriers in China.

IV. Strategic and Tactical Analysis

Armed with the positional analysis of markets, firm competencies, and the nonmarket environment in different geographical contexts, firms are positioned to undertake a set of strategies and implement them to succeed in Asian market. Because strategic and tactical analyses are deeply interrelated and frequently overlap in the case studies, we can examine these two aspects together in discussing our findings.

Strategic and Tactical Analysis: Review of the Theory

To review briefly, strategic analysis refers to how firms respond to and attempt to manipulate market forces. Efforts to develop market strategies have been analyzed from many perspectives. Particularly helpful is the work of Richard D'Aveni, who argues that firms compete in four different *arenas:* cost and quality, timing and know-how, strongholds, and deep pockets.[6] In the cost and quality arena, firms begin with a homogenous product and compete for market share through price differentiation. As price wars escalate, they must shift their focus to quality and service to gain market share. Timing and know-how refer to the ability of firms to seize control of the market, based on such classic concepts as first mover advantages (or disadvantages) and the innovative character of their products. Strongholds are crucial because they may provide firms with the ability to exclude competitors from particular regional, industrial, or product segments. But in a dynamically changing market, such barriers are likely to erode quickly and firms must seek new strongholds. Finally, the deep pockets arena focuses on the ability of firms to utilize their superior financial resources to discourage weaker competitors.

Organizational strategy considers how firms restructure to compete in light of their positional analysis and choice of market and nonmarket strategy. While this is not a central focus of the volume, key issues include how to organize to compete in trade and investment, based on considerations of transaction cost factors. For example, with respect to investment, should firms attempt to create wholly owned subsidiaries or would a minority owned operation suffice? Market forces and strategies will clearly affect this decision. For example, might a minority owned operation use the home firm's technology to become a competitor down the line? Often, however, a more critical question concerns the nonmarket environment in which firms are operating, including political hazards.

At the level of nonmarket strategy, firms must engage in calculations about their possible supporters and opponents on issues of critical importance for success. These include questions about the demand side (what benefits will different actors receive from success on an issue) and on the supply side (who will be able to generate political action?). These considerations will often influence a firm's decision on market strategy. VW, for example, aggressively pursued local suppliers in China, in large part to ensure it would receive political support from key officials in China.

Turning to tactics, firms must assess their abilities to execute market and nonmarket strategies and build competencies in this area as needed. Market tactics refer to firms' decisions regarding R&D, production, and marketing as they strive to compete in various market arenas. Organizational tactics involve the internal restructuring of their management and organizational structure. Finally, nonmarket tactics concern policies that might be pursued to advance both market and nonmarket strategies. These include lobbying, grassroots activity, coalition building, testimony, political entrepreneurship, electoral support, communication and public advocacy, and judicial strategies.[7]

The Case Studies:
Devising and Implementing Strategies for Asia

How have firms positioned themselves strategically to win in Asian markets? In the software industry, European firms that had been positioned to compete in national European markets quickly found themselves at a severe competitive disadvantage with the advent of the PC. As a result, U.S. producers were the first to dominate the rapidly growing packaged software segment, which places a premium on the development of low cost, high quality products with short product cycles. The link to hardware producers proved to be a crucial advantage in this market segment, and network externalities reinforced this lead. However, Nakagawa's research suggests that the proliferation of PCs in corporate Europe created a new market opportunity in enterprise resource planning and inter-enterprise software where European local expertise could be leveraged into a comparative advantage. This new demand allowed firms such as SAP and BAAN to leverage their strength in devising customized solutions and services by creating complex standardized packages for multiple industrial users.

To implement their strategy, these firms successfully built alliances in Asia with local and global partners, provided technical training to customers through developing partner academies, and willingly personalized their global solutions across sectors. All of these tactics played to the strength of European firms who had expertise in service and meeting customized demands. A second key niche in Europe arose from the complexity of languages, organizational cultures, and differences in legal and accounting systems. Here, too, firms were able to build on their previous competencies in meeting the needs of large firms. These opportunities have helped to drive large firms such as Olivetti and Groupe Bull to undertake radical restructuring and reorganization to focus on services, components, and telecommunications. Although other European multinationals were the first clientele for these firms, many firms like Sterling and Micro Focus chose to focus on producing packaged software for a global market niche. Their ability to succeed in Asia, however, still remains an open question as suggested by the example of Synon and recent demise of BAAN in late 2000.

From a nonmarket perspective, many European software firms have utilized national and EU funding and support to develop reusable and standardized software products. As noted, however, this has not easily translated into success in Asia. Still, the policies of many Asian governments toward IT industries in general has made for a favorable market and nonmarket environment. This has led to a more nurturing environment for firms with tax breaks, relocation incentives, pre-competitive research grants, and assistance in locating and training qualified local personnel. The variation in openness can be seen in the success that European and other nationals' firms have had in Taiwan and Singapore—which sought to become regional hubs—as opposed to the difficulty they have faced in penetrating markets in Japan and South Korea. In Japan, for example, SAP has successfully advanced its market penetration through a joint venture with Fujitsu and other European firms as an entrée into other Asian countries. European firms that have succeeded in Asia have recognized, however, that a U.S. presence is often necessary. Thus, as Nakagawa notes, in many cases the consolidation of activities and alliances is an essential component of successful entry into Asian software markets.

The lucrative but highly competitive Chinese market provides a key battleground for auto firms. Although by the mid-1990s the Japanese had made inroads in most of Asia, China was still a relatively untouched market. The key challenge for all firms was to develop both market and nonmarket strategies in view of the many nonmarket hazards in China. To understand this integrated strategy, Biziouras and Crawford compare the entry and relative degrees of success of two key European auto firms, Peugeot and Volkswagen, in the Chinese market. Both VW and Peugeot entered the Chinese market in the mid-1980s with joint venture arrangements with Chinese manufacturers, and both firms signed additional agreements in the early 1990s. By the end of that decade, however, VW had emerged much more successful.

With respect to market strategy and tactics, several key decisions gave VW a significant edge over Peugeot. First, based on previous experience in Europe and Latin America, VW worked actively to set up an extensive local production network of suppliers in China. While this was in response to the nonmarket constraints of state restrictions on imports, this strategy had important market benefits as well. It gave VW flexibility in responding to market demands, and allowed it to meet pressures for domestic content. Second, VW created a broad distribution network of 400 centers, creating an important stronghold against competitors. Moreover, its service network of two hundred service stations—as compared to Peugeot's of less than one hundred—allowed VW to develop a reputation for good customer service. Third, VW introduced new models in response to changing demand, while Peugeot was slow to react. Fourth, VW hired Chinese managers to work with its German managers in pairs, while Peugeot relied on expatriate French managers. VW also created training institutes for workers and promoted its best workers into management positions. Fifth, VW developed a significant R&D facility to produce cars with the latest technology that met market demand in Asia, with the objective of using its Chinese production facilities to export to Southeast Asia.

VW recognized the need to pursue simultaneously an active nonmarket strategy in view of the complexity of the Chinese state. In marked contrast to Peugeot, VW executives were in China constantly. Moreover, VW relied heavily on the German federal government as well as support at the state (*Laender*) level. VW's development of local production and creation of an R&D center also increased its influence with regional and central government officials. In view of the ongoing decentralization of economic decisionmaking power to regions, VW's close ties to Shanghai officials proved critical in securing support, both political and financial, in the form of loans and a host of preferential policies.

Unlike European software firms, the European aerospace consortium Airbus successfully competes head-to-head with an American firm, Boeing, for market share in Asia. Sandholtz and Love discuss a number of specific market strategies that have given Airbus a strong regional presence. Its imitation of Boeing's "family concept"—using the same wings, fuselage, and/or cockpits for many plane designs and making relatively smaller production adjustments related to length and size—has allowed it to reduce production costs. Indeed, Airbus has gone even farther than Boeing in this direction. This strategy has also helped it increase market share due to the greater number of products available, the speed with which different planes can be delivered, and the reduced training requirements for pilots. More-over, some of its aircraft products such as the A330/340 allow it to meet both the demand for thin, long haul markets and high-volume regional service. Airbus has also excelled at introducing advanced technology, in-cluding fly-by-wire, cockpit automation, and composite materials. In ad-dition, Airbus has also bested Boeing with respect to delivery time, although the latter is now becoming increasingly competitive with shorter production lead times. On pricing, Airbus has often undercut Boeing, lead-ing the latter to complain about government subsidies to Airbus.

Airbus has fared less well on other dimensions. While it initially suc-ceeded in displacing sales of the MD-11 because of its range problems, its A340 now faces trouble from the Boeing 777 because of the A340's slower speed, narrower cabin, and inability to meet range guarantees. The 777 has also given Boeing an advantage in responding to Asian demands for high average seating capacity, and its decision to enlist three Asian airlines in de-signing the 777 has stood it in good stead. To meet this challenge, Airbus has moved forward with the A380, a double decker jumbo launched in late 2000. It has also moved to improve its after-sales service, which lagged be-hind Boeing in the cost of spare parts and speed of delivery. Finally, with respect to operating costs, Boeing and Airbus appear to be tied. For exam-ple, while its A340 is seen to have lower operating costs than the 777, the opposite is true with its A330.

Although Airbus would not have succeeded in Asia without a strong market strategy, it has complemented this effort with a variety of nonmar-ket strategies and tactics. One key approach has been the use of subcon-tracting and co-production. With activist Asian states seeking technology transfers and jobs, and with state owned airlines as customers, such arrangements have been an essential element of Airbus's success. Examples include the technical advice given to the Indonesians on their turboprop commuter plane, the subcontracting of components for Airbus planes, and training centers in several Asian countries. Still, these efforts have not al-ways guaranteed orders, and collaborative efforts have sometimes fallen through. As a consortium of nationally-owned European aerospace com-panies, Airbus has benefited from diplomatic pressure on the part of their home governments to encourage Airbus purchases. Strategies include the offer of landing rights for Asian airlines, concessionary financing, and un-related links to trade concessions. The most controversial issue has been launch aid from home governments. After considerable conflict with Boe-ing, an agreement was reached in 1992 to limit direct government support to 33 percent of the total cost of programs. Airbus and Boeing, backed by their home governments, have both accused one another of market interference through bribes and government pressure.

Finally, Wallner examines Asian banking and insurance markets. As noted in the positional analysis, nonmarket factors in the form of regulation have strongly limited the breadth of the market, making products relatively homogenous. This key market characteristic reduces firms' space for strategic maneuvering. In many cases, then, smaller and medium-sized firms have limited their activities to meeting the needs of firms from their own countries who have invested in Asia. The exception has been in the banking industry, where firms have engaged in a combination of market and nonmarket strategies to curry favor with large firms and government related entities that would otherwise not be profitable. The objective of these strategies has been to develop relationships with an eye to developing long-term investment banking or bond flotation business. In some cases, they have also pursued successful organizational tactics such as strategic alliances. In 1997, for example, the Swiss Bank Corporation allied with Long-term Credit of Japan to develop a common range of services in the Japanese market. From a nonmarket perspective, Wallner's research suggests that European banks faced much tougher going, because they did not have government backers from the EU with the same political clout as those of Japanese or American firms.

In his case study of the Japanese casualty insurance industry, Wallner finds that European firms have fared significantly worse than American firms, accounting for a meager 0.3 percent versus 2.4 percent share, respectively. The latter have used pressure from the U.S. government to press for market liberalization and have been able to make inroads into the market. Moreover, once having entered, American firms have successfully prevented other foreign firms from encroaching on their market segments.

V. Strategic Lessons

Firms attempting to penetrate Asian markets, either through trade or foreign direct investment, have faced significant market and nonmarket obstacles. In the sectors that we have examined, firms responded by using a variety of market, nonmarket, and organizational strategies. In many cases, similar strategies were applied successfully in various sectors; in others, their effectiveness has been limited by specific sector characteristics. In this section, I discuss general lessons from the case studies and suggest directions for future research.

Figure 9.1 Market Challenges and Strategies in Asia

	Strategies	
	Market	Nonmarket
Market challenges (in italics)	*High start-up costs:* Cost sharing with partners, subcontracting, and co-production.	*High start-up costs:* Home government subsidies, relocation incentives, tax breaks, pre-competitive research grants.
	Barriers to entry: Early and aggressive entry using allies and joint ventures; build distribution network to create barriers.	*Barriers to entry:* Host and home government promotion efforts; liberalization and transparency efforts.
	Market responsiveness: Local R&D to product differentiate; local suppliers and managers; local customization of product; utilize core competencies but go beyond existing skill set.	*Market responsiveness:* Trade and investment liberalization, R&D subsidies, and tax incentives.
	Network externalities: Locate in the U.S. and other centers of industry concentration.	*Network externalities:* Firm-government cooperative efforts to share training costs of public sector employees and adoption of new products.
	Economies of scale: Become a multinational corporation by locating in the U.S. market as well as in Asia.	*Economies of scale:* Host and home government promotion programs.

To get a sense of the types of generic strategies that appear to be successful in Asia, we must consider two types of challenges faced by firms: market and nonmarket. As we have seen, firms respond to each by undertaking both market and nonmarket strategies. We can combine these two categories of problems and responses (see figures 9.1 and 9.2) to generalize about the types of responses that seem effective. Figure 9.1 reviews the individual market and nonmarket strategies that have proven successful in responding to market conditions across various sectors.

Responses to Market Challenges

Examination of strategic problems in Asia typically focuses on the market problems and opportunities posed by rapid economic and population growth. Analysts have advised firms to seek economies of scale, divide production globally according to labor costs, and the like. But the possibility of using nonmarket responses to gain market share and profits in Asia has been given relatively short shrift. Yet firms that focus only on market strategies have frequently been outmaneuvered by their competitors. Without an integrated market and nonmarket strategy, firms will not adequately respond to market and nonmarket challenges.

High Start-Up Costs. Firms in many sectors have faced high start-up costs. The most extreme example of this is in the aerospace industry, but firms in the automobile, software, and financial sectors have also had to cope with this challenge. In the cases of Airbus, Volkswagen, and SAP, effective market responses included cost sharing efforts with partners, subcontracting, and co-production. For example, by shifting risk to local producers as well as devolving production regionally, VW achieved significant market presence. This strategy also proved effective in the nonmarket sense. Nonmarket strategies include lobbying for home government subsidies and aid, as in the case of Airbus and several European software companies, or lobbying for financial assistance, tax breaks, and other incentives from host countries. For example, VW benefited from its government relations in Germany and China, where it secured a key position as a privileged firm, giving it access to state assistance and loans, as well as guarantees with respect to entry by other foreign firms into the Chinese market.

Barriers to Entry. Barriers to entry and aggressive competition have been common challenges for firms attempting to succeed in Asia. The most effective firms have addressed these issues by seeking allies and creating joint venture operations. VW, Airbus, SAP, and European financial firms have all employed such strategies with success. VW and SAP were particularly aggressive in creating partnerships with local firms. VW went on to build an impressive network of distributors and service stations that created a significant barrier to entry for other firms.

In this area, nonmarket strategies have proven especially effective. For example, Airbus has been able to use home country lobbying to secure purchases of its aircraft, with home governments linking these purchases to landing rights, technical assistance, and concessionary finance. Similarly, VW effectively cultivated the support of the German government, whereas the efforts of Peugeot fell short. This difference is all the more striking in light of the efforts of the European Commission to develop common policies. It is also worth noting that the lobbying by European governments, as compared to that of others, has not always borne equal fruit. In the insurance industry, for example, lobbying by European governments was considerably weaker than that of the American and Japanese governments on behalf of their firms. In many cases, of course, Asian host countries have attempted to attract investment by liberalizing and increasing transparency—even without prodding from an investor's home government.

Market Responsiveness.   Firms that were willing to adapt their products to the demands of local Asian markets using local R&D and product differentiation were particularly successful. Moreover, the use of local suppliers— through subcontracting, for instance—combined with the hiring and promotion of local managers gave firms such as VW a significant advantage over its competitors. Software firms such as SAP and Synon have succeeded in Asia by customizing packaged products for large enterprises and working with local firms and governments. SAP has created a ready supply of skilled labor by establishing training centers in Asia, which allow it to better service its customers. As compared to other European software firms, which have relied on existing competitive advantages in services and skill in customization based on their previous experiences in Europe, SAP has been willing to look beyond its existing core competency that had been developed in the home market. From a nonmarket perspective, firms have been able to secure or take benefit from subsidies and tax incentives to improve their market responsiveness. For example, the EU helped to subsidize research in software as noted with respect to start-up costs, and host governments have attempted to entice firms in software to locate in Asia through a host of incentives. In many cases, firms may be able to lobby for various types of financial aid (as in VW's case in China) to enhance their competitive position.

Network Externalities.   In some industries, such as software and autos, the penetration of Asian markets depends on developing novel designs and new products. In an environment with network externalities, simply locating in low cost areas or where the company has been located for historical reasons may be insufficient. For this reason, European software firms have located elements of their operations in Silicon Valley or other areas with high concentrations of firms that would stimulate innovation, encourage partnerships, and give them access to capital markets. Similarly, European auto firms have located design operations in Los Angeles to benefit from the high degree of expertise in that region. Although responsiveness to local market and nonmarket demands is a crucial element of firm strategy in Asia, locating in the presence of other firms to encourage the stimulation of network externalities can also be important for overall global and regional success. Firm efforts to secure aid from governments to encourage training of their employees or adoption of new products has also been an effective nonmarket strategy. For example, by working with the Singaporean and Taiwanese governments, SAP has been able to expand demand for its products and create externalities by promoting adoption of their products by governments and large enterprises alike.

Economies of Scale.   With costs declining rapidly in response to large scale production in many industries such as software, autos, and aircraft, developing a global market for products is essential. Thus, from a market perspective, many software firms have sought to locate in Silicon Valley, not only to benefit from network externalities, but to secure business customers to increase the scale of their production. This issue is obviously central in aircraft, where the high capital costs and falling costs with longer production runs makes it imperative to sell globally. Thus, Airbus has sought large numbers of customers before committing to the A380. Similarly, in autos, longer production runs significantly reduce costs, giving VW a significant incentive to use China as a production base to sell more widely in Southeast Asia. From a nonmarket perspective, VW, Airbus, and SAP were able to benefit from various EU, European government programs, as well as Asian host promotional programs, thus allowing them to achieve economies of scale to reduce costs and make them more competitive with other firms. VW, for example, received tax exemptions, policy-oriented loans, and priority in using foreign funds and listing in stock and bond markets, which allowed it to build larger plants and achieve scale economies.

Figure 9.2 Nonmarket Challenges and Strategies in Asia

	Strategies	
	Market	Nonmarket
Nonmarket challenges (in italics)	*Local content requirement:* Local R&D operations; use local suppliers and local managers to meet requirements.	*Local content requirement:* Lobby home and host governments to renegotiate; join host country's regional accords; push for sectoral or multilateral accords.
	Highly regulated market: Local alliances, including joint ventures and locally owned subsidies.	*Highly regulated market:* Organizational experience in other markets to build nonmarket competency; lobbying by home government; coalition with local suppliers.
	Subsidies: Establish foreign-owned subsidiary or develop local partnership.	*Subsidies:* Lobby host and home to create new rules to access funds or declare such subsidies unfair.
	Protectionism: Local strategic alliances; use local suppliers and distributors.	*Protectionism:* Lobby home and host countries to implement further liberalization under WTO; push for sectoral, bilateral or regional accords.
	Standards consortia: Join local consortia directly or indirectly through local partners.	*Standards consortia:* Obtain home and host country support to adopt your technology standard.

Responses to Nonmarket Challenges

In Asia, state intervention in the market is the norm—rather than the exception. While the so-called East Asian model has been criticized in the aftermath of the economic crises, this propensity of governments to manage markets has not disappeared. Thus, firms must be able to respond to and lobby governments, regional organizations, and even the World Trade Organization (WTO) as a crucial element of their strategy. Figure 9.2 illustrates the strategic responses to nonmarket challenges.

Local Content Requirements.   Many countries have imposed local content requirements on firms, forcing them to source locally for a portion of their production. Free trade agreements also generally impose some kind of regional content requirement to prevent firms from simply exporting goods to a country with low barriers and then shipping them to the partner country with higher barriers to avoid duties or other market impediments. In response to these requirements, VW in China developed local R&D facilities and encouraged the use of local managers. In that case, this strategy was an effective response to both market and nonmarket concerns. As noted above, local managers and production increased VW's capacity for market responsiveness. From a nonmarket perspective, firms can also lobby governments to alter the domestic content requirements imposed by governments or regional arrangements. They might also push for participation in regional arrangements such as the Association of Southeast Asian Nations (ASEAN) members auto accords. More globally, firms in the software industry have been particularly active in promoting the development of the information technology agreement (ITA) to liberalize trade in these products. European firms have also taken an interest in developments in the Asia–Pacific Economic Cooperation forum (APEC), as well as pushing their governments to actively participate in arrangements such as the Asia-Europe Meetings (ASEM) to keep abreast of and maintain control over these developments.

Highly Regulated Markets.   Although local content requirements can be met relatively easily with market responses, other types of barriers are often less amenable to market strategy. In such cases, nonmarket strategies, particularly the lobbying of host governments and help from home governments, may be required. Firms with experience in dealing with activist governments or authoritarian regimes may have developed significant competencies to deal with regimes such as the highly interventionist Chinese government. VW had been involved in producing in Eastern Europe for some time, and was well prepared to address this type of nonmarket concern. VW was also adept at cooperating with its local suppliers to lobby for aid and other benefits, a strategy Airbus has also used in Asian markets.

When faced with nonmarket constraints, firms in the insurance, auto, and aerospace industries have all sought help from their home governments. However, in the insurance industry, European firms appeared to have fared worse then their Japanese and American counterparts because of the greater leverage the U.S. and Japanese governments have in China. By contrast, in the auto and aerospace sectors, VW and Airbus succeeded in attracting significant assistance in both locating plants and in selling products.

Subsidies. As we have seen, host countries in Asia have been highly interested in securing investment in the software, auto, and aircraft sectors. European firms such as SAP, VW, and Airbus have all actively developed local partnerships to benefit from such aid. By working with local firms that are eligible for such subsidies, these firms have been able to advance several market and nonmarket objectives simultaneously. Subsidies by home countries also have been helpful. Synon, for example, became an innovator in the design of computer-aided design engineering (CASE) tools. Its global strategy benefited from favorable UK science and technology tax policies as well as European development funds. Such home country subsidies have, of course, been very controversial at times, leading to sharp conflict among governments. The best example is the ongoing conflict between Airbus and Boeing that has engaged the United States and EU to define limits on start-up aid and efforts to subsidize production and sales.

Protectionism. Despite ongoing liberalization, many Asian markets restrict access in many areas. Firms often are not able to export directly to these countries, or even secure needed parts. In the case of autos and the financial services industry, we have seen how firms have coped with such protectionism by engaging in foreign direct investment. Thus, VW and Peugeot recognized the need to directly produce in China. VW's local sourcing strategy allowed it to overcome import restraints, and proved useful in addressing a host of other challenges. In the banking sector, European banks sought to penetrate protected markets for financial services by engaging in nonprofitable or high-risk bank lending to develop goodwill with potential customers. Nonmarket strategies with respect to protection have been particularly significant. In software, firms were active in pushing the ITA agreement through—first by lobbying actively in APEC, and then in the WTO. In autos, VW used its privileged position to discourage liberalization by China, and its management board even called for a delay in China's accession to the WTO unless it was granted safeguards to protect domestic infant industries such as the auto industry.

Standards Consortia. In sectors where standards are particularly important, such as the software, banking, and insurance industries, the setting of standards can prove to be a key determinant of market success. In software, involvement in standards consortia has been an essential market strategy. Partnering with firms in Asia served SAP in good stead, and SAP was also able to work directly with Asian governments in various joint research centers that could be instrumental in setting standards. Other European firms have found that unless they locate in the United States and "become more American," they may face the danger of being left out of the development of new standards. From a nonmarket perspective, in cases where home industry was in danger of being left behind by increasing foreign firm market share (as in the software industry), the European Commission has taken the initiative to encourage a consortium of firms. Its objective has been to avoid duplication of effort, provide venture capital funds and to encourage the development of common standards in order to help European firms position themselves vis-à-vis their American competitors. As we have seen, however, these efforts in dynamic sectors have not always produced the desired result.

Future Avenues for Research

This book has provided a framework for examining how firms can succeed in Asian markets. Positional analysis has given us insight into the diversity of different sectoral environments, and a focus on strategic responses to market and nonmarket challenges has provided us with insight into winning strategies for Asia. The broad set of sectors covered by the case studies provides a spectrum of experiences from which we can draw in generalizing about optimal market and nonmarket strategies.

What types of research and further investigation will allow us to better understand the development of firm strategies for Asian markets? First, while the case studies have given us insight into comparative strategy, the focus of this book has been the experiences of European firms. More detailed analysis of the strategies employed by American and Japanese firms in penetrating Asian markets would help us to advance our understanding of the role played by different home governments as well as the unique characteristics of firms of different nationalities.

Second, while our analysis has considered a range of sectors, it would be useful to consider what patterns might hold and what new lessons might emerge from studies of the behavior of firms in other areas. For example, do firms in the telecom or chemical industries—both of which are highly regulated by governments—pursue significantly different strategies than firms in electronics or autos?

Third, firm strategies may change over time, in response to both changing market environments as a result of the Asian crises and to evolving bilateral, regional, and international arrangements. Will progressive liberalization in Asia as a response to the recent crises and as a result of pressures from international financial institutions create a significantly different regional environment? These questions have been important to our analysis, because we have been able to examine firm strategies before and after the Asian crises. Still, while many changes continue to take place in Asia, we submit that the long-term implications of these changes are better understood and predicted using the integrated market-nonmarket framework we present in this volume.

Two companion volumes to this book—*Winning in Asia, American Style* and *Winning in Asia, Japanese Style*—which focus on the strategies of both American and Japanese firms, will address these ongoing questions. In view of the complexity of business-government interaction in Asia, however, this field should remain a fertile ground for years to come.

Notes

1. As noted in the introduction, two companion volumes systematically explore the strategies of American and Japanese firms in the region.
2. See also Aggarwal and Morrison (1998) for a discussion of the weakness of APEC decisionmaking apparatus.
3. Porter (1980).
4. Hamel and Prahalad (1994).
5. Baron (1999, 2000).
6. D'Aveni (1994).
7. See Baron (1999, 2000) for discussion of these nonmarket tactics.

References

Aggarwal, Vinod K. and Charles Morrison, eds. (1998). *Asia-Pacific Crossroads: Regime Creation and the Future of APEC.* New York: St. Martin's Press.

Baron, David (1999). "Integrated Market and Nonmarket Strategies in Client and Interest Group Politics," *Business and Politics* 1 (1) (April).

Baron, David (2000). *Business and Its Environment,* 3rd edition. Upper Saddle River, NJ: Prentice Hall.

D'Aveni, R. (1994). *Hypercompetition: Managing the Dynamics of Strategic Maneuvering.* New York: The Free Press.

Hamel, Gary and C. K. Prahalad (1994). *Competing for the Future.* Boston: Harvard University Press.

Porter, Michael E. (1980). *Competitive Strategy.* New York: Free Press.

Index

* numbers in bold are for tables and figures

ABB, 97
Abraham, Reinhardt, 204
Aérospace Matra, 198
Aerospatiale, 97, 188, 193, 197, 207
Aggarwal, Vinod, 60, 170
AIG, 249
Air Canada, 215
Air France, 188, 211
Air India, 204, 213, 217
Airbus, 20, 25, 187–218, 263–264, 268, 269, 272, 273, 274, 276–277
Airbus China, 215
    see Airbus Integrated Company
Airbus Consortium
    see Airbus
    see Airbus Integrated Company
Airbus Industrie (AI),
    see Airbus
    see Airbus Integrated Company
Airbus Integrated Company (AIC), 198
Alcatel, 97, 145
Alenia, 210
All Nippon Airways (ANA) 200, 202, 203, 208
Altshom, 97
American Motor Corporation (AMC), 182
Apple, 134
Argentina, 229
Asia-Europe Business Conference, 73
Asia-Europe Business Forum, 20, 73, 95
Asia-Europe Meeting (ASEM), 20, 23, 52, 60, 62, 72–73, 95, 161, 259, 260, 276
    history of, 52, 72–73
    comparison with APEC, 72–73

Investment Promotion Action Plan, 73, 95
Trade Facilitation Action Plan, 73, 95
Asian Financial Crisis, 4, 31–32, 59–60, 225, 232, 236, 257, 275
    Asian responses to, 59–60, 232
    economic context and history, 31–32
Asia-Pacific Economic Cooperation (APEC), 4, 7–9, 20, 23, 48, 50, 60–62, 64, 69–72, 74, 260, 276–277
    and International Technology Agreement, 4, 9
    Bogor Summit, 70
    comparison with ASEAN, 71
    history of, 69–72
    APEC Business Advisory Council (ABAC), 71–72
Association of European Automobile Constructors, 98
Association of Petrochemical Producers (APPE), 87, 98
Association of Manufacturers of Insulated Wires and Cables (EUROPACABLE), 98
Association of Southeast Asian Nations (ASEAN), 7–8, 20, 23, 31, 35, **37–39, 40–41,** 46, 48, 51, 60–69, 72–73, 161, 260, 276
    and European Investment, 23, **37–39**
    ASEAN Automotive Federation, 63
    ASEAN Chambers of Commerce and Industry (CCI), 64, 67, 69, 260
    ASEAN Federation of Textile Industries, 67

ASEAN Free Trade Area (AFTA), 48,
    65–66, 260
ASEAN Industrial Complementation, 63
ASEAN Industrial Cooperation, 69
ASEAN Industrial Joint Venture
    Agreement, (AIJV) 64–65
Basic Agreement on Industrial Projects,
    63
Common Effective Preferential Tariff
    (CEPT), 65, 67
contrast with NAFTA, 66
history of, 62–69
Junior Managers Program, 51
Preferential Trading Arrangements
    (PTA), 64–65
widening and deepening, 64–65
Australia, 8, 62, 141, 201
Austria, 166

BAAN, 128, 129, 138, 265, 266
Balladur, Edouard, 214
Bank of China, 172, 173, 181
Banque Nationale de Paris, 172
Baron, David, 12, 13, 135, 262
Basic Agreement on ASEAN Industrial
    Projects, 63
BASF, 46
Bayer, 97
Beijing Jeep Corporation, 169, 180
Berne Convention, 121
Biziouras, Nick, 24, 263, 267
BMW, 97
Boeing, 20, 25, 187–191, 193–194,
    196–199, 204–205, 207, 210,
    212–213, 216, 218, 263–264, 268,
    269
Boisso, Dale, 47
Bosnia, 214
Bouyges, 97
Brazil, 229
Britain, *see* United Kingdom
British Aerospace, 97, 209
British Aircraft Corporation, 193, 197
British Petroleum (BP), 97
Brown, Adam, 203, 209, 211
Brunei, 67
Bundestag, 178
Buscelhofer, Robert, 162

CA/Sterling, 111
Calvet, Jacques, 177
Canada, 166, 215
Cap Gemini, 126, 129, 137, 140, 151
Cathay Pacific, 199, 201–205, 207
CCCL, 147
Chile, 229
China, 20–21, 24, 31, 35, **37, 39, 40–41,**
    46, 60, 61, 64, **113, 117, 118,** 119,
    121, **122,** 131, 160–162, 164–165,
    167–183, 209–210, 213–214, 226,
    228, 232, **238,** 263–264, 266–267,
    272, 274, 276–277
Chinese Community Party (CCP), 167
Civil Aviation Authority of China (CAAC),
    215,
China Airlines, 212
China Aviation Supply Corporation
    (CASC), 210, 215
China International Trust and Investment
    Corporation (CITIC), 172, 180
China National Automotive Industry, 180
Chirac, Jacques, 178
Cho, Yang Ho, 203
Churchill, Winston, 59
Clinton, William, 214
Closer Economic Relations (CER), 8, 62
Coface, 173
Cold War, 60, 168
Comité Consultatif International
    Téléphonique et Télégraphique
    (CCITT), 20
Common Effective Preferential Tariff
    (CEPT), 65, 67
Computer Associates, 112
    see CA/Sterling
Construcciones Aeronáuticas (CASA), 197
Crawford, Beverly, 24, 263, 267
Credit Suisse, 97
CSC, 130

D'Aveni, Richard, 14, 16, 133, 135, 265,
    216
Daimler-Benz, 97, 213, 214
DaimlerChrysler, 97, 165, 169, 263
DaimlerChrysler Aerospace (DASA), 193,
    197, 198
Danone, 98

Debis, 129
Deutsche Airbus, 188, 193, 213
Deutsche Bank AG, 237
Douglas
  see McDonnell Douglas
Dragonair, 200
Dupont, Cédric, 23, 60, 258, 260, 261

East Asian Economic Caucus, 61
East Asian Economic Grouping, 61
East Asian Economic Model, 258–259, 275
Eastern Airlines, 217
Economic and Monetary Union (EMU),
  32, 83, 90
Economic Growth
  general post War, 31, 258
  Europe compared with Asia (1980s),
    31–32
  Asian prospects, 53–54, 161–163, 225,
    263
Economist, The, 246
EDS, 126
Elf, 97
Ericsson, 97, 98
European Association of Manufacturers of
  Business Machines and
  Information Industry
  (EUROBIT), 98
European Central Bank, 82
European Chemical Industry Council
  (CEFIC), 98
European Commission, 8, 64, 68, 72, 79,
  81, 83, 85, 87, 90, 93–94, 96, 260,
  273, 278
European Community (EC), 20, 32, 46, 79,
  208, 213
  European Community Investment
    Partners (EPIC), 46, 52
European Conference of Associations of
  Telecommuncations (ECTEL),
  98
European Council, 81, 83, 85, 90, 93, 94
European Court of Justice, 81
European Economic Community, 80
European Investment Bank (EIB), 52
European Parliament, 74, 79, 81, 83, 85, 214
European Round Table of Industrialists
  (ERT), 87, 93

European Systems and Software Initiative
  (ESSI), 124
European Union (EU), 7, 24, 32, **37,**
  40–41, 42–43, 46, 48, 51–52,
  61–62, 69, 72, 74, 79–99, 108,
  159, 161, 169, 215, 248, 258–261,
  263, 267, 270, 273, 277
  ASEM Business Forum, 20, 73, 95
  Asia-Invest Program, 46, 52, 96
  Bureau de Rapprochement des
    Enterprises (BRE), 96
  Business Cooperation Network (BC-
    NET), 96
  Commissioner for External Relations,
    95
  Committee of Permanent
    Representatives (COREPER), 84
  Council of Ministers, 81, 261
  Council of Foreign Affairs, 85
  Directorate-General for Enterprise
    Policy, Distributive Trades,
    Tourism, and Cooperatives
    (DGXXIII), 96
  Directorate-General for External
    Economic Relations (DGI), 95
  Directorate-General for Science,
    Research, and Development
    (XII), 95–96
  European Strategic Program for
    Research and Development in
    Information Technology
    (ESPRIT), 96
  Investment Promotion Action Plan
    (IPAP), 73, 95
  Junior Managers Program, 51
  Lobbying avenues and strategies, 79–99
  Research and Technological
    Development Policy, 96
  see Maastricht Treaty
  see Treaty of Rome
  Trade Facilitation Action Plan (TFAP),
    73, 95
  widening and deepening, 51
Evans, Gareth, 62
Export-Import Bank, 208

Fahd, King, 214
FAW-VW Automotive Company, 172

Ferrantino, Michael, 47
Fiat, 97
Finland, 165
Fire and Marine Rating Insurance
        Association of Japan, 242
Five Forces Model, 10, 163, 164, 234, 262
Ford, 165, 166
    Ford China, 162
    Ford Europe, 97
Foreign Direct Investment, 23, 31–36,
        38–54, 59, 64, 65, 167–168, 234,
        258, 259, 270
    Asian attitudes toward, 59
    Chinese policy toward, 167–168
    compared with portfolio investment,
        35
    European history in Asia, 32, 38–47,
        53–54
    European prospects in Asia, 53–54
    history in Asia prior to 1997, 33–35
    intra-Asian, 39–46, 259
    overview in Asia after 1997, 35
    Relative liberalization compared to
        other financial services, 225
France, 46, 98, **113, 115,** 173, 178, 197,
        211–213, 229, **238–240**
Fuji, 209
Fuji Heavy Industries, 165–166
Fujitsu, **110,** 142

Gandeler, Bruno, 179
Garuda Indonesia, 209
GEC Alsthom, 46
General Agreement on Tariffs and Trade
        (GATT), 8–9, 48, 64, 70, 94, 208,
        213
    GATT Agreement on Trade in Civil
        Aircraft, 213, 214
    see World Trade Organization (WTO)
    Uruguay Round, 8, 64, 70
General Motors Europe, 97
General Motors Shanghai, 164
Germany, 46, 98, 108, **110, 113, 115,**
        140–141, 166, 168, 170, 176–178,
        183, 193, 197, 229, 237, **238–240,**
        268, 272–273
Group of Five (G5), 34

Groupe Bull, 137, 145, 266
Guangzhou Automobile Manufacturing
        Firm, 171
Guangzhou Peugeot, 179
Gucci, 98
Guizhou Aviation, 209

Hahn, Karl, 170, 177
Halliday, George, 168
Hamel, Gary, 12, 262
Hashimoto, Prime Minster, 245
Hawker Siddeley, 193
Hermès, 98
Hinisz, Witold, 17
Hoffman-La Roche, 97
Holland, **238–240**
Hong, Fang, 176, 177
Hong Kong, **37, 39,** 61, **113, 117–118,**
        120, 122, 131, 226, 228, 230, **238,**
        258, 200, 201, 213
Hewlett Packard, **110**
HSBC Holdings, 97

Iberdrola, 97
IBM, 108, **110,** 126, 129, 130, 134, 143,
        144, 147
IBM Germany (Germany), 140
ICI, 97
ING Group, 97
India, **117–118,** 119, **122,** 131, 161–162,
        165, 204, 212, 217
Indonesia, 23, 31, **37, 39, 42–43,** 46, 60,
        62, 63, **113, 117–118,** 121, **122,**
        131, 161–162, 165, 184, 209, 229,
        238, 258
Industrial Sectors
    Aerospace, 8, 9, 24, 94, 97, 99, 257, 263,
        264, 268–270, 187–218
    case studies, 187–218
    Agriculture, 8, 9, **38, 42–45**
    Automotive, 9, 20, 21, 25, **38,** 63, 68,
        94, 97, 159–184, 257, 267–268,
    case studies, 159–184
    Chemicals, 9, **38,** 46, 94, 97, 98, 99
    Electronics, 8, 9
    Energy, 9, 94, 99
    Environmental, 9

Financial Services, 8–9, 20, 24, 25, **43,**
45, 46, 94, 96, 97, 98, 99,
225–250, 257, 264, 270
banking industry, 226, 228,
235–237, 257, 270
case studies, 225–250
insurance industry, 226, 227, 228,
237, 241–242, 257, 270
securities industry, 226
Fisheries/Aquaculture, 9, 70
Machinery and Transport Equipment,
36, **38, 42–45,** 97
Mining, **38, 42–45**
Oil and Petroleum, **42–45,** 87, 97–98
Paper, **42–43**
Software, Information-Technology, 24,
94, 96, 97, 98, 99, 107–151, 257,
265
case studies, 135–148
Steel, 8, 11, **38**
Telecommunications, **38, 43, 45,** 96, 97,
98
Textiles, **38, 42–45,** 67
Tobacco, **38, 42–45,** 46
Information Technology Agreement (ITA),
4, 9, 123, 276, 277
Institutions, International
deepening, 8, 51, 64–65
widening, 8, 51, 64–65
see Asia-Europe Business Forum
see Asia-Europe Meeting
see Asia-Pacific Economic Cooperation
see Association of Southeast Asian
Nations
see European Union
see General Agreement on Tariffs and
Trade
see Group of Five
see International Monetary Fund
see North American Free Trade
Association
see Organization for Economic
Cooperation and Development
see United Nations
see World Bank
see World Trade Organization
Intel, 114

International Air Transport Agreement
(IATA), 194
International Finance Corporation, 172
International Monetary Fund (IMF), 14,
33, 48, 49, 232, 259
1997, Financial Crisis, 33, 232
International Telecommunications Union
(ITU), 4, 20
Intersolv, 144, 146, 149
Investment
see Foreign Direct Investment
see Portfolio Investment
Investment Promotion Action Plan (IPAP),
73, 95
Isuzu, 165
Italy, **110, 113, 115, 238–240,** 210

Jaegar, 211
Japan, 9, 20–22, 31–33, **37, 40–41,** 50, 54,
61, 64, 97, **110, 113–115,** 116,
117, 119, **120, 122,** 125,
131–132, 134, 142–143, 159–162,
165–166, 169–171, 202, 208–209,
213, 225, 227–231, 233–234, **236,**
237, **238–240,** 241–243, 245–250,
263–264, 267, 270, 273, 277–278
Insurance Business Law, 241
Liberal Democratic Party (LDP), 247,
Ministry of Finance (MOF), 241–246,
249–250
Official Development Assistance
(ODA), 50
amakudari, 244
keiretsu, 233, 242, 250
Japan Airlines (JAL), 202, 203
Japan Atomic Energy Insurance Pool, 242
Japanese Hull Insurers' Union, 242
Jennings, David, 202, 203
John, Stewart M., 201
Jung, Ku-Hyun, 73

Kawai, Masahiro, 47
Kawasaki, 209
Kinkel, Klaus, 177
Kohl, Helmut, 177, 215
Kong, Cheong Choong, 202, 204
Korean Air Lines (KAL) 194, 203, 212

Koshkarian,Vaughn, 162
Krupp 97
Kuwait Airlines 194

Lafarge, 97
Laos, 228
Lehmann, Jean-Pierre, 73
Lloyds TSB, 97
Lobbying, 8, 22, 69, 79–99, 260–261
    ASEAN countries, 69
    European Strategies in European
        Union, 98–99
    European Union, 79–99, 85–96,
        260–261
    Typology of Strategies, **89**
Lockheed, 193, 196, 217
Logica, 137
Long-Term Credit Bank of Japan, 233, 270
Love, William, 25, 263, 268
Lufthansa, 188
Luxembourg, 90
LVMH, 98

Maastricht Treaty, 51
Magna, 166
Malaysia, 31, **37, 39, 42–43,** 46, 61–62,
        67–69, **113, 117–118,** 121, **122,**
        131, 148, 161–162, 165, 184, 201,
        229, **239,** 258
Malaysia Airlines System (MAS), 202, 209
Malaysia, Federation of, 62
Market Forces
    see positional analysis
    see strategic analysis
    see tactical analysis
Mazda, 165
McDonnell Douglas, 191, 193, 197, 207,
        210, 213, 217
MERCOSUR, 68
Merant, 111, 112, 145, 146, 149
Messerschmitt-Beolkow-Blohm (MBB),
        193
Mexico, 229
Michot, Yves, 207
Micro Focus, 111, 112, 131, 266
Microfocus, 24
Microsoft, 108, **110,** 112, 114, 126, 138,
        144, 147

Middle East Airlines, 194
Ministry of Finance (MOF), 241–246,
        249–250
    and Anti-Monopoly Law, 242
Mitsubishi, 209
Mitsubishi Motors, 165, 169
Mitterand, François, 178–179, 214
    see Merant
    case study, 142–147
Mohamad, Mahathir, 61
Montinola, Gabriela, 167
Mowery, David, 191

Nakagawa, Trevor H., 24, 262, 263
National Academy of Engineering, 192
National Research Council, 192
Naughton, Barry, 167
NEC, **110**
Nestlé, 93–94, 98
Netscape, 129
New Zealand, 8
Newly Industrializing Countries (NICs), 4,
        31, 34–35, **37, 39, 40–41,** 46, 258
Newly Industrializing Economies, (NIEs)
    see Newly Industrializing Countries
Nippon Life, 234
Nokia, 97, 98
Nonmarket Forces
    see positional analysis
    see strategic analysis
    see tactical analysis
North American Free Trade Association
        (NAFTA), **37,** 62, 66
    contrast with AFTA, 66
North Korea, 60
North Vietnam, 60
Northwest Airlines, 204
Novartis, 97
Novell, 126

Official Development Assistance (ODA), 50
Olivetti, **110,** 137, 266
Opel, 166
Oracle, 126, 131
Organization for Economic Cooperation
        and Development (OECD), 94,
        130
Organizational Pressures

see positional analysis
see strategic analysis
see tactical analysis

Pacific Economic Cooperation Conference, 72
Pakistan, 206
Pakistan International Airlines, 205
Pan Am, 213
Pardoe, Alan, 200
Peoplesoft, 131
Peruda, 164
Petrofina, 97
Peugeot, 24, 97, 160, 161, 169, 171, 173–182, 266, 267, 268, 277
    Guangzhou Peugeot, 180
Philippines, **37, 39,** 46, **113, 117–118, 122,** 131, 199, 228, **239**
Philippines Airlines, 199
Philips, 93–94
Pierce, J. B. L., 212
Pierson, Jean, 204, 211
Plattner, Hasso, 141
Porsche, 166
Porter, Michael, 7, 10, 16, 163, 234, 262
    Five Forces Model, 10, 163, 164, 234, 262
Portfolio Investment, 65
Positional Analysis, 5–14, 111–125, 161–170, 187–197, 225–250, 261–264
    Case study: Airbus vs. Boeing in Asia, 187–197
        Firm Position, 189–196
        Market Position, 189–196
        Nonmarket Position, 196–197
    Case study: Banking Services in Asia, 225–250
        Firm Position, 228–230, 235–237
        Market Position, 226, 228–230, 235–237
        Nonmarket Position, 231–233, 235–237
    Case study: European Auto Firms in China, 161–170
        Firm Position, 168–170
        Market Position, 163–167
        Nonmarket Position, 167–168

Case study: European Software Industry, 111–125
    Market Position, 112–116
    Nonmarket Position, 116–125
Case study: Insurance Industry in Asia, 225–250
    Firm Position, 228–230, 237–248
    Market Position, 226, 228–230, 237–248
    Nonmarket Position, 231–233, 237–248
Definition, 5–14
Firm Position, 6, 7
Graphically Illustrated, **6–7**
Market Position, 6, 7
Nonmarket Position, 6, 7
Posth, Martin, 175, 179
Prahalad, C. K., 12, 262
Preferential Trading Arrangements (PTA), 64, 65
Prisoner's Dilemma, 244–245
Proton, 164
Proximity Software, 146
Putnam, 234

Qian, Yingyi, 167

Rolls Royce, 214
Ravenhill, John, 23, 258, 259, 260
Renault, 94, 97
    Renault-Nissan, 166
Rhône-Poulenc, 97
Rongji, Zhu, 178, 179
Rosenberg, Nathan, 191
Rue, Rolf, 210
Russia, 59, 69, 235
Ryan, William F., 208

Saint-Gobain, 97
Salzer, Bernard, 214
Sandholtz, Wayne, 25, 263, 268
SAP AG, 108, **110,** 111, 126, 129, 130, 133, 134, 138–142, 146, 265, 266, 272, 273, 277, 278
    case study, 139–142
SAP Labs, 140
Saudi Arabia, 194
Saudi Arabia Airlines, 194

Schrempp, Jurgen, 166
Sears, 22
Sema Group, 137
Sfena, 211
SGS-Thomson, 139
Shanghai Auto Industry Corporation, 173
Shanghai Automobile, 180
Shanghai Tractor and Automobile
        Corporation, 171
Shell, 97
Shenyang Aircraft Corporation, 209
Shirk, Susan, 167
Siemens AG, 97, 98, 134, 142, 145
Siemens-Nixdorf, 114
Simultan, 129
Singapore, 31, **37, 39,** 46, 62–63, 67, 69, 73,
        113, 117–118, 119, **120, 122,** 131,
        134, 140, 143, 199, 210, 211, 226,
        228, 230, **239–240,** 258, 267, 274
Singapore Airlines (SIA), 199, 201, 202,
        203, 204, 207
Singapore Declaration, 66, 67
Single European Act (SEA), 81
Single European Market (SEM), 81
Sogema, 211
Solnick, Steven, 168
South Korea, 20, 31, **37, 39, 42–43,** 46,
        59–61, **113, 117–118, 120, 122,**
        131, 134, 143, 161–162, 165, 169,
        194, 209, 225–226, 228–230,
        232–233, 237, **238–239,** 250, 258,
        264, 267
    chaebols, 233, 237, 250
South Vietnam, 60
Sovereignty, 60
Soviet Union, 168
Soros, George, 234
Spain, 124, 197
Specification and Programming
        Environment for Communication
        Software (SPECS), 124
Spero, Joan, 71
Sterling Software, 112, 144, 266
    see CA/Sterling
Steyr-Dailmer-Puch, 166
Strategic Analysis, 5, 14–20, 86–88, 91–94,
        125–135, 170–184, 197–215,
        225–250, 264–266, 269

Case study: Airbus vs. Boeing in Asia,
        197–215
    Firm Strategy, 197–208
    Market Strategy, 197–208
    Nonmarket Strategy, 208–215
Case study: Banking Services in Asia,
        225–250
    Market Strategy, 234–235, 235–237
    Nonmarket Strategy, 234–235,
        235–237
    Organizational Strategy, 233–234,
        235–237
    Overview, 233, 235–237
Case study: European Auto Firms in
        China, 170–184
    Market Strategy, 173–177
    Nonmarket Strategy, 177–182
    Organizational Strategy, 172–173
    Overview, 170–172
Case study: European Software Industry,
        125–135
    Market Tactics, 125–132
    Nonmarket Tactics, 132–135
Case study: Insurance Industry in Asia,
        225–250
    Market Strategy, 234–235,
        237–248
    Nonmarket Strategy, 234–235,
        237–248
    Organizational Strategy, 233–234,
        237–248
    Overview, 233, 237–248
    Definition, 5, 14–20
    Graphically illustrated, **6**
    Market Strategy, 6, 265, 269
    Nonmarket Strategy, 6, 265, 269
        of the European Union, 86–88,
        91–94
    Organizational Strategy, 6, 265
Stuart, Colin A., 199
Sukarno, President, 62
Sun Microsystems, 126
Sutch, Peter, 204
Suzuki Motors, 165, 166
Swiss Bank Corporation, 233, 270
Synon, 111, 112, 131, 144, 149, 266, 273,
        274, 277
    see CA/Sterling

case study, 147–148

Tactical Analysis, 5, 21–23, 125–135,
     170–184, 197–215, 225–250,
     264–266
  Case study: Airbus vs. Boeing in Asia,
     197–215
     Firm Tactics, 197–208
     Market Tactics, 197–208
     Nonmarket Tactics, 208–215
  Case study: Banking Services in Asia,
     225–250
     Market Tactics, 234–235, 235–237
     Nonmarket Tactics, 234–235,
        235–237
     Organizational Tactics, 233–234,
        235–237
     Overview, 233, 235–237
  Case study: European Auto Firms in
     China, 170–184
     Market Tactics, 173–177
     Nonmarket Tactics, 177–182
     Organizational Tactics, 172–173
     Overview, 170–172
  Case study: European Software Industry,
     125–135
     Market Tactics, 125–132
     Nonmarket Tactics, 132–135
  Case study: Insurance Industry in Asia,
     225–250
     Market Tactics, 234–235, 237–248
     Nonmarket Tactics, 234–235,
        237–248
     Organizational Tactics, 233–234,
        237–248
     Overview, 233, 237–248
  Definition, 5, 21–23
  Graphically illustrated, **6**
  Market Tactics, 6, 21, 265, 269
  Nonmarket Tactics, 6, 22–23, 265, 269
  Organizational Tactics, 6, 21, 265
Taiwan, 31, **37, 39,** 60–61, **113, 117,** 119,
     120, 122, 131, 134, 162, 165,
     178–179, 213, 225, 228–229, **240,**
     258, 267, 274
Thailand, 31–32, **37, 39,** 46, 63, **113,**
     117–118, 121, **122,** 131, 162, 184,
     207, 211, 213, 228–229, **240,** 258

Thai International Airlines, 207, 213
Thai Airways, 211
Thiessen, 97
Tiananmen Square, 179
Tractor Industrial Corporation, 180
Toa Domestic Airline (TDA), 208
Tokyo Stock Exchange, 241
Total, 97
Trade, International, 31, 33, **37,** 234, 258,
     259, 270
  Asian share of, 33
  European-Asian, **37**
  Intra-Asian, 48, 65, 68
  relationship with FDI, 48
Trade Facilitation Action Plan, 73
Trade Related Investment Measures
     Agreement (TRIMS), 8
Treaty of the European Union (1991), 81
  see Maastricht Treaty
Treaty of Rome, 80–81, 95
Tyson, Laura D'Andrea, 191, 192

UBS, 97
Unilever, 98
Union of Industrial and Employers'
     Confederation (UNICE), 98
Union of Machinery Insurers of Japan, 242
United Kingdom, 62, **113,** 142–4, 193, 208,
     213, 217, 229, 230, **238–240,** 241,
     277
  pressuring Asia for liberalization, 4, 20
United Nations, 48
United Nations Conference on Trade and
     Development (UNCTAD), 48,
     259
United States of America, 4, 20–21, 23, 32,
     37, 40–41, 50, 61, 107, 109, **110,**
     112, **113–115,** 119, **120,** 123, 125,
     131–132, 136, 141–143, 145–147,
     150, 159, 165–166, 183, 197, 201,
     206, 208–209, 212–214, 217–218,
     227, 229–230, 235, **236,** 237,
     238–240, 241, 245, 247–249, 262,
     270, 273, 277–278
  Department of Defense 197
  Office of Technology Assessment,
     (OTA) 191
  Pentagon, 196–197

United States-Japan Insurance Agreement,
        248
United States Trade Representatives, 245
Urata, Shujiro, 23, 47, 258, 259

Valmet, 166
Vietnam, 121, **122,** 228, **240,** 211
Vietnam Airlines, 211
Vivendi, 97
Volkswagen (VW), 24, 97, 160–162,
        166–168, 169–178, 181–184,
        263, 265–267, 272–274, 276–277
    Shanghai VW, 174, 175
Volvo, 160
von Tein, Volker, 203

Wallner, Klaus, 25, 259, 264, 270
Waters, Martin, 145, 146

Weingast, Barry, 167
Welkener, Burkhard, 178
White, Gorden, 168
Wilde, Clarence V., 212
Williamson, Oliver, 16
World Bank, 33, 48, 49, 236
    and 1997 Financial Crisis, 33
World Intellectual Property Organization
        (WIPO), 121
World Trade Organization, 4, 8, 61, 67, 70,
        74, 94, 134, 163, 226, 276, 277
    and ITA, 4
    compared with APEC, 70
World War II, 229, 241

Xian Aircraft Corporation, 209, 210

Zemin, Jiang, 178, 179, 211

Printed in the United States
By Bookmasters

Printed in the United States
By Bookmasters